T0135503

Sterols in *Daphnia* nutrition:
physiological and ecological consequences

Dominik Martin-Creuzburg

Bibliografische Information Der Deutschen Bibliothek

Die Deutsche Bibliothek verzeichnet diese Publikation in der Deutschen
Nationalbibliografie; detaillierte bibliografische Daten sind im Internet über
http://dnb.ddb.de abrufbar.

ISBN 3-8325-1179-2

Logos Verlag Berlin
Comeniushof, Gubener Str. 47,
10243 Berlin
Tel.: +49 030 42 85 10 90
Fax: +49 030 42 85 10 92
INTERNET: http://www.logos-verlag.de

Dissertation der Universität Konstanz

Tag der mündlichen Prüfung: 5. August 2005

Referent: PD Dr. E. von Elert

Referent: Prof. Dr. K.-O. Rothhaupt

Referent: Prof. Dr. K. Dettner

Table of contents

Chapter 1

General Introduction

Nutritional constraints at the phyto- zooplankton interface—In recent years, trophic interactions between phyto- and zooplankton have been in the focus of attention, mainly because the efficiency of energy transfer across the plant-herbivore interface is highly variable with often far-reaching consequences for the performance of higher trophic levels. Cyanobacterial carbon, for instance, is inefficiently transferred to higher trophic levels, which often leads to a decoupling of primary and secondary production in eutrophic lakes where cyanobacteria dominate the phytoplankton (De Bernardi and Giussani 1990). But what are the factors that regulate the observed variations in the efficiency with which carbon is transferred to primary consumers? This question led to the food quality concept which implies that not all units of food are equally nutritious for their consumers (reviewed by Sterner and Schulz 1998). The topic of the present work is the identification of nutritional constraints at the phytoplankton—zooplankton interface by measuring food quality effects on zooplankton performance. In freshwater ecosystems, cladocerans, in particular the genus *Daphnia*, are key organisms which provide a crucial link between primary and secondary production. Daphnids, however, are non-selective suspension feeders and, thus, should be particularly sensitive to the dominance of nutritionally inadequate food particles. Therefore, the identification of nutritional constraints between phytoplankton and the herbivore *Daphnia* might be essential to gain insight into the factors that regulate the energy transfer in freshwater food webs.

Nutritional inadequacy can be attributed to numerous factors, first of all to toxicity or morphological properties which prevent ingestion or digestion. Attention has also been drawn to the elemental or biochemical composition of food organisms. Among the elements, phosphorus, which is often poorly available in freshwater ecosystems, has been shown to be potentially limiting for zooplankton growth. Daphnids are believed to be limited by phosphorus when their food source exceeds a molar carbon-to-phosphorus ratio (C:P) of approximately 300 (Sterner and Schulz 1998). However, it has been shown that, if phophorus is available in amounts adequate to meet the demands, the biochemical composition of the food has to be considered. Animals show a dietary need for a large set of biochemical nutrients, such as certain vitamins, amino acids and fatty acids. Recent attention has focused on long-chain polyunsaturated fatty acids (PUFAs), which play important roles in animal physiology (Cook 1996). The absence of certain PUFAs has been proposed as a potential food quality constraint (Müller-Navarra 1995, Müller-Navarra et al. 2000, Wacker and Von Elert 2001, Von Elert 2002, Becker and Boersma 2003, Ravet et al. 2003). More recently, the dietary sterol content has been identified as another crucial parameter in determining food quality for *Daphnia* (Von Elert et al. 2003).

In the field, biochemical characteristics of the seston are determined by its species composition. Phytoplankton diversity is often highly variable, subject to seasonal fluctuations and depends on the trophic state of the ecosystem. The seston of eutrophic lakes is often dominated by cyanobacteria. This is equivalent to a low availability of essential biochemical nutrients, since these are often hardly represented in cyanobacteria. In general, cyanobacteria do not contain long-chain PUFAs (Ahlgren et al. 1992) and, as prokaryotes, do not synthesize sterols (Volkman 2003). This led to the hypothesis that the low carbon transfer efficiency at the cyanobacteria—*Daphnia* interface, leading to the accumulation of cyanobacterial biomass, is due to the lack of either of these essential lipid classes (DeMott and Müller-Navarra 1997, Von Elert and Wolffrom 2001, Von Elert et al. 2003). However, according to Liebig's law of the minimum, only one factor can actually be limiting at a certain point of time (Von Liebig 1855). Von Elert et al. (2003) provided evidence that trophic interactions between cyanobacteria and the herbivore *Daphnia* are primarily constrained by the lack of sterols and that PUFAs become limiting only when the shortage of sterols had been overcome by sterol supplementation. This suggests that the absence of sterols has to be considered as the major food quality constraint of cyanobacteria.

Physiological properties of sterols in arthropods—The inability to synthesize sterols de novo is a characteristic feature of all arthropods, which distinguishes them from most other animals. In eukaryotic organisms sterols are essential for several physiological processes and, therefore, arthropods must acquire these nutrients from their diet. Cholesterol tends to be the principle sterol in animals, although small amounts of dietary sterols are frequently found, in particular in herbivorous species. Behmer and Nes (2003) recently proposed that sterols play at least three key roles in arthropod physiology: (i) as structural components of cell membranes; (ii) as precursors to ecdysteroid hormones; and (iii) as signaling molecules bound to hedgehog proteins. Sterols are indispensable structural components of eukaryotic membranes. Their amphipathic nature enables the incorporation into phospholipid bilayers: the polar head group (3β-OH) oriented to the aqueous phase and the non-polar sterol nucleus and the alkyl side chain located in the hydrophobic core of the phospholipid bilayer, interacting with the fatty acyl chains (Fig. 1). Their capacity to modulate the membranes fluidity is propably the most striking role of sterols in eukaryotes (Nes and McKean 1977).

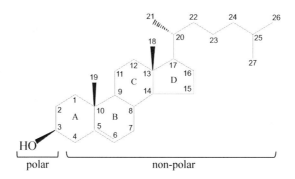

Fig. 1 Cholesterol

Furthermore, sterols serve as precursors for the ecdysteroids which are involved in the process of molting. In general, ecdysone and subsequently 20-hydroxyecdysone, the two most active ecdysteroids in arthropods, are synthesized from cholesterol (Grieneisen 1994, Gilbert et al. 2002; Fig. 2). In crustaceans, ecdysteroids are produced mainly in steroidogenic glands, the so called Y-organs. Our knowledge of ecdysteroid biosynthesis in crustaceans others than decapods is scarce.

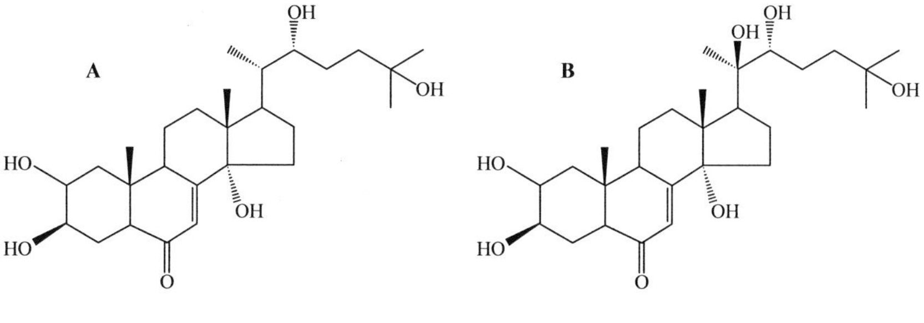

Fig. 2 Ecdyson (A) and 20-hydroxyecdyson (B)

7

Recently, cholesterol has been found to covalently modify proteins of the hedgehog family. Hedgehog proteins are secreted signaling proteins that are required for developmental patterning of various embryonic structures in insects and other organisms (Porter et al. 1996). These findings support the multiple role of sterols in eukaryotes and open up a new field of interest.

Life-history consequences of sterol availability—Considering the dietary need of sterols in arthropods together with the key positions sterols occupy in animals physiology, it is not surprising that the dietary sterol content is a crucial parameter which determines the performance of arthropods in their environment. The importance of sterol availability to growth and developmental processes has long been recognized. However, most studies have focused on insect nutrition (reviewed by Behmer and Nes 2003), and our knowledge of the role of dietary sterols in crustaceans is poor and mostly restricted to decapod crustaceans, which have received attention due to their relevance for aquaculture. In chapter 2, I investigated possible effects of a reduced sterol availability on life history traits of *Daphnia*.

Cholesterol, the most important sterol in animal physiology, is the predominant sterol found in arthropods and in most other animals (Goad 1981). Thus, carnivorous species are readily supplied with cholesterol. Herbivorous species, however, cannot rely on a dietary source of cholesterol since this sterol is hardly represented in plant material. Instead, plants and algae contain several types of phytosterols that differ from cholesterol by additional substituents (e.g. methyl or ethyl groups at C-24) or by the position and/or number of double bonds in the side chain or in the sterol nucleus (Piironen et al. 2000, Moreau et al. 2002). Herbivores can either use the sterols present in their diet directly or they have to metabolize them to cholesterol to meet the requirements for growth and development (Svoboda and Thompson 1985). Apart from a few exceptions, herbivorous insects and, so far examined, also crustaceans use dietary sterols to synthesize cholesterol. Thereby, most species studied are capable of dealkylating and reducing common C-24-alkyl phytosterols, such as sitosterol or stigmasterol, to cholesterol (Grieneisen 1994, Behmer and Nes 2003; Fig. 3). However, more than 200 different types of phytosterols have been reported in plant material (Moreau et al. 2002), and not all of them are suitable as cholesterol precursors. Numerous studies have focused on dietary needs for sterols in insects and revealed that the pattern of sterol metabolism is by no means ubiquitous (Behmer and Nes 2003). In crustaceans, comparable studies are scarce and, referring to cladocerans, they are not available at all. Cladocerans are unable to select particles individually with regard to their quality. Therefore, the predominance of dietary sterols which are unsuitable to serve as cholesterol precursors might have serious life history consequences; an aspect which is pesented in chapter 3.

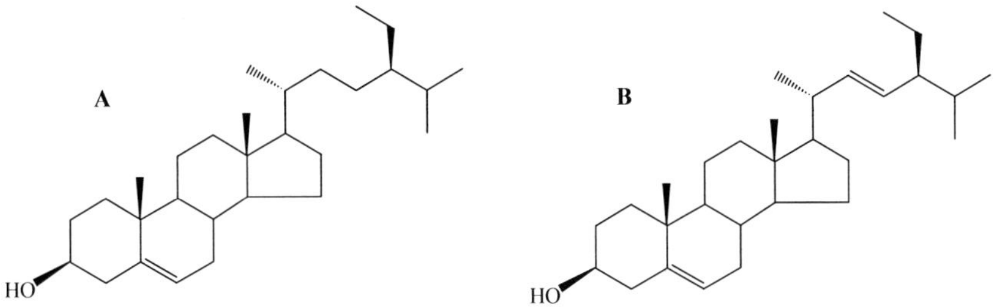

Fig. 3 Sitosterol (A) and stigmasterol (B)

Trophic upgrading of autotrophic picoplankton food quality by protozoans—In oligo- to mesotrophic lakes, autotrophic picoplankton (APP) is a major player in carbon production, thereby forming the base of a complex microbial food web (Weisse 1993, Callieri and Stockner 2002). Heterotrophic nanoflagellates (HNFs) and small ciliates significantly prey upon unicellular picocyanobacteria, the major constituents of APP. As described above, cyanobacteria are a poor quality food source for cladocerans, mainly due to their biochemical composition. Heterotrophic protists as intermediary grazers might improve the biochemical mediated food quality of picocyanobacteria for subsequent use by higher trophic levels (e.g. *Daphnia*). Experimental evidence exists that heterotrophic flagellates as intermediary grazers are capable of biochemically upgrading a poor quality food source for crustacean zooplankton by producing essential lipids (Klein Breteler et al. 1999, Bec et al. 2003, Tang and Taal 2005). Thereby, most attention has focused on PUFAs supplied by the intermediary protozoans, although Klein Breteler et al. (1999) already emphasized the potential role of sterols for the observed trophic upgrading of a PUFA- and sterol-deficient diet. In chapter 4, the freshwater heterotrophic flagellate *Paraphysomonas* sp. was used to investigate if protists as intermediary grazers are capable of upgrading an APP food source for subsequent use by *Daphnia* by producing essential lipids such as PUFAs and sterols.

The capability of ciliates as intermediary grazers to upgrade a poor quality food source has received far less attention. Recently, Klein Breteler et al. (2004) did not find evidence for trophic upgrading by a marine ciliate for copepods. Ciliates are often poor in PUFAs that are essential for *Daphnia* and do not synthesize sterols de novo (Conner 1968, Harvey and

McManus 1991, Harvey et al. 1997, Klein Breteler et al. 2004); hence, trophic upgrading by ciliates appears uncertain. Instead of sterols, however, ciliates produce the pentacyclic triterpenoid alcohol tetrahymanol and related compounds (e.g. diplopterol, Fig. 4), which provide functional equivalence to sterols as structural components of cell membranes (Conner et al. 1968). In chapter 5, I assessed the role of ciliates in transferring mass and energy from APP- to *Daphnia*-production and linked possible effects on the performance of the herbivore to tetrahymanol and related compounds provided by the ciliates.

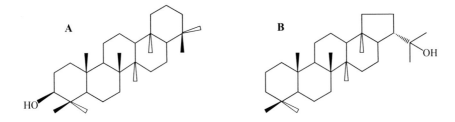

Fig. 4 Tetrahymanol (A) and diplopterol (B)

Available data suggest that ciliates are less nutritious for daphnids than many algae (DeBiase et al. 1990, Wickham et al. 1993, Sanders et al. 1996). In chapter 6, we addressed the question whether this nutritional inadequacy of ciliates can be attributed to the lack of sterol in these protists. Although ciliates are incapable of synthesizing sterols de novo, exogenously supplied sterols can be incorporated into cell membranes and metabolized into various other sterols (Conner et al. 1968, Harvey and McManus 1991, Harvey et al. 1997). Since sterols might differ in their suitability to support *Daphnia* growth, these structural changes might also affect the food quality of ciliates for *Daphnia*.

Chapter 2

Life history consequences of sterol availability in the aquatic keystone species *Daphnia*

Dominik Martin-Creuzburg, Alexander Wacker and Eric von Elert

Abstract—The absence of essential biochemical nutrients, such as polyunsaturated fatty acids or sterols, has been considered as a mechanism determining trophic interactions between the herbivore *Daphnia* and its phytoplankton food source. Here we experimentally quantify the sensitivity of two *Daphnia* species to decreasing amounts of dietary sterols by measuring variations in life history traits. The two species *D. magna* and *D. galeata* were fed different mixtures of the sterol-containing green alga *Scenedesmus obliquus* and the sterol-free cyanobacterium *Synechococcus elongatus*; a higher proportion of *Synechococcus* in the food is equivalent to a decrease in dietary sterols. To address the significance of sterol limitation, the *Daphnia* species were also fed *Synechococcus* supplemented with cholesterol. In both species, somatic and population growth rates, maternal dry mass, the number of viable offspring, and the probability of survival were significantly reduced with the lower availability of sterols. A high correlation between the sterol content of the mixed diet and the somatic and population growth rates was found, and growth on cholesterol-supplemented *Synechococcus* fitted well into this correlation. Somatic growth of first-clutch neonates grown on 100% *Synechococcus* exhibited a pattern similar to that of somatic growth of their mothers grown on the different food regimes, which demonstrated the significance of maternal effects for sterol-limited population growth. *D. galeata* had a two-fold higher incipient limiting sterol level than *D. magna*, which indicated interspecific differences in sterol requirements between the two *Daphnia* species. The results suggest a strong impact of dietary sterols on life history traits and therefore population dynamics of the keystone species *Daphnia*.

Keywords—food quality, sterol limitation, maternal effects, population growth, survival

11

Introduction

The flow of energy in aquatic food webs is often constrained by the low transfer efficiency of organic carbon to higher trophic levels. In particular, interactions between primary producers and their herbivorous consumers can lead to a reduced productivity at the first consumer level and to a decoupling of primary and secondary production. Crustacean zooplankton of the genus *Daphnia* are key organisms in aquatic ecosystems; they are effective filter feeders with high grazing impact on the phytoplankton and serve as the most important link between primary production and the production of many fish species (Lampert 1987). Daphnids feed on a broad range of particle sizes and are unable to select particles individually with regard to their quality.

Cyanobacteria are known to be an inadequate food source for freshwater cladocerans (De Bernardi and Giussani 1990), but the reasons determining this dietary inadequacy are controversial. Grazing resistance attributed to morphological properties, toxicity, and the absence of essential elemental or biochemical nutrients are considered to be responsible for the lack of grazer control on cyanobacterial assemblages, leading to the accumulation of cyanobacterial biomass and therefore to reduced water quality (Carmichael 1994). In freshwater ecosystems, the available phosphorus (P) is often scarce, and daphnids have been shown to be P-limited when their food source exceeds a carbon-to-phosphorus ratio (C:P) of approximately 300 (Sterner and Hessen 1994; DeMott 1998). However, C:P ratios <300 are found in the seston of many lakes (Brett et al. 2000), and in these cases, other factors appear to be potentially limiting for *Daphnia* growth (Sundbom and Vrede 1997).

Polyunsaturated fatty acids (PUFAs) are essential biochemical nutrients, and a low availability of single PUFAs in natural seston (Müller-Navarra 1995; Müller-Navarra et al. 2000; Wacker and von Elert 2001) and in monoalgal food (von Elert 2002; Ravet et al. 2003) has been shown to constrain growth of *Daphnia*. The fact that cyanobacteria do not contain long-chain PUFAs (Ahlgren et al. 1992) has led to the hypothesis that the low carbon transfer efficiency at the cyanobacteria-*Daphnia* interface is due to the lack of PUFAs. However, von Elert and Wolffrom (2001) have suggested that the absence of a non-PUFA lipid present in eukaryotic algae might be responsible for the poor assimilation of cyanobacterial carbon.

Sterols are another class of lipids that are absent in most prokaryotic cyanobacteria (see Volkman 2003 for a critical review), and recently von Elert et al. (2003) have shown that the absence of sterols can constrain the carbon transfer between cyanobacteria and the herbivore *Daphnia galeata*. Like insects, crustaceans are incapable of synthesizing sterols de novo and, therefore, must acquire these essential nutrients from their diet (Goad 1981). Cholesterol, the principal sterol in crustaceans, is an indispensable structural component of cell membranes and serves as a precursor of steroid hormones, such as ecdysteroids, which are involved in the process of molting (Goad 1981). Unlike carnivorous crustacea, the herbivorous daphnids cannot rely on a dietary source of cholesterol since only trace amounts are found in phytoplankton species (Nes and McKean 1977). However, eukaryotic phytoplankton is often rich in various phytosterols, some of which are suitable precursors for the synthesis of cholesterol (Martin-Creuzburg and von Elert 2004).

The focus of the present study was to quantify the sensitivity of the freshwater cladoceran *Daphnia* to decreasing amounts of dietary sterols by measuring variations in life history traits. For this purpose, two species of daphnids (*D. magna* and *D. galeata*) were fed different mixtures of the sterol-containing green alga *Scenedesmus obliquus* and the sterol-free cyanobacterium *Synechococcus elongatus*. Thus, a higher proportion of *Synechococcus* in the food is equivalent to a decrease in dietary sterols. To address the significance of sterols for proper growth, the two *Daphnia* species were also fed *Synechococcus* supplemented with cholesterol. The coccoid *Synechococcus elongatus* was chosen because it is a nontoxic, phosphorus-rich cyanobacterium that is well-assimilated by *Daphnia* (Lampert 1977*a*,*b*; 1981), but is deficient in essential fatty acids (von Elert and Wolffrom 2001) and lacks sterols (Martin-Creuzburg and von Elert 2004). In addition to the determination of juvenile somatic growth rates, which are often used as a relative measure of fitness in *Daphnia* (Lampert and Trubetskova 1996), we applied a demographic approach to estimate the intrinsic rate of population increase (r) and its dependence on the dietary sterol content. Furthermore, effects of maternal lipid investment on the juvenile somatic growth rates of the first-clutch offspring were investigated.

Materials and methods

Cultivation and preparation of food organisms—The green alga *Scenedesmus obliquus* (SAG 276-3a, Sammlung von Algenkulturen Göttingen, Germany) was used as food for the stock cultures of the daphnids. It was grown in batch cultures in Cyano medium (Jüttner et al. 1983) and harvested in the late-exponential growth phase. For growth experiments, *Scenedesmus obliquus* and the cyanobacterium *Synechococcus elongatus* (SAG 89.79) were each cultured semi-continuously in Cyano medium (20°C; illumination at 120 and 60 μ mol $m^{-2} s^{-1}$, respectively) at a dilution rate of 0.25 d^{-1} in aerated 5-l vessels; the cells from 25% of the culture were harvested every day, replacing the sampled medium with freshly prepared medium. Stock solutions of these food organisms were obtained by centrifugation and resuspension in fresh medium. The carbon concentrations of the food suspensions were estimated from photometric light extinction (800 nm) and from carbon-extinction equations determined previously.

For simplification, the following abbreviations will be used for the different food regimes: 'Scen' for *Scenedesmus obliquus* and 'Syn' for *Synechococcus elongatus*. Five food suspensions containing different proportions of *Scenedesmus* and *Synechococcus* were prepared. The total carbon concentration of 2 mg C l^{-1} was represented by 100% Scen, 80% Scen + 20% Syn, 50% Scen + 50% Syn, 20% Scen + 80% Syn, and 100% Syn. In a sixth treatment, *Synechococcus* was supplemented with cholesterol (Sigma). For the supplementation, 30 mg bovine serum albumin (BSA) was dissolved in 7.5 ml of ultra-pure water, and 600 μl of an ethanolic stock solution of cholesterol (2.5 mg ml^{-1}) was added. Subsequently, the solution was brought to 60 ml with 6 mg particulate organic carbon (POC) of the *Synechococcus* suspension and Cyano medium. After incubation on a rotary shaker (100 revolutions min^{-1}) for 4 h, excess BSA and free cholesterol were removed by washing the cells three times in fresh medium according to von Elert (2002). The resulting suspension was used as food in the growth experiments. Preliminary experiments showed that BSA did not affect growth of *Daphnia*.

Growth experiments—Growth experiments were conducted with third-clutch juveniles (birth ± 6 h) of a clone of *Daphnia magna* (originally isolated from Großer Binnensee, Lampert 1991) and a clone of *Daphnia galeata* (originally isolated from Lake Constance, Stich and Lampert 1984). The experiments were carried out at 20°C in glass beakers filled with 1-l of filtered lake water (0.45-μm pore-sized membrane filter). Each treatment consisted of three replicates with 20 (*D. magna*) or 55 (*D. galeata*) animals per beaker. The food suspensions were renewed daily until the animals released their third clutch. At day 6 of the experimental period, 5 *D. magna* or 10 *D. galeata* individuals were sub-sampled, dried for 24 h, and weighed on an electronic balance (Mettler UMT 2; ±0.1 μg). The juvenile somatic growth rates (g) were determined as the increase in dry mass from the beginning of an experiment (W_0) to day 6 (W_t) using the equation:

$$g = \frac{\ln W_t - \ln W_0}{t}$$

The remaining animals were kept in the corresponding treatments until they reached maturity. Shortly before the release of the juveniles, 5 *D. magna* or 15 *D. galeata* individuals were transferred separately to 100 ml jars without adding food. After juveniles hatched, the dry mass of the mother and of the neonates was determined (within 0–3 h after release from the brood pouch). The same procedure was used for the second and third clutch. In order to estimate maternal effects of the different food regimes on the performance of the offspring, subsamples of first-clutch neonates were reared 6 days on 100% *Synechococcus* (sterol-free food conditions), and the juvenile somatic growth rates were determined as described above. Population growth rates (r) were estimated iteratively using the Euler-Lotka equation:

$$1 = \sum_{x=0}^{n} l_x \, m_x \, e^{-rx},$$

where l_x is the age-specific survivorship, m_x is the age-specific fecundity (number of neonates per individual), and x is the age at reproduction (in days). The probability of survival until reproduction (l_x) was estimated from the mortality that occurred in the different treatments. Growth rates were calculated as the means of each treatment.

Analyses—Sterols were extracted from approximately 0.5 mg POC of the food suspensions according to Martin-Creuzburg and von Elert (2004). Free sterols were quantified with a gas chromatograph (HP 6890, Agilent Technologies, Waldbronn, Germany) equipped with an HP-5 capillary column (Agilent) and a flame ionization detector by comparison with an internal standard (5α-cholestan). Sterols were identified using a gas chromatograph-mass spectrometer (Finnigan MAT GCQ) equipped with a fused silica capillary column (DB-5MS, J&W); the instrumental settings are described elsewhere (Martin-Creuzburg and von Elert 2004). Mass spectra were identified by comparison with mass spectra of a self-generated spectra library or mass spectra found in the literature. POC was determined with an NCS-2500 analyzer (ThermoQuest GmbH, Egelsbach, Germany).

Statistical analyses—Somatic and population growth rates of *Daphnia* were analyzed using 1-way analyses of variance (ANOVA). The dry mass of mothers and neonates, and the number of offspring were analyzed in full-factorial designs. The experimental factors were either food category (1-way) or food category and number of successive clutches (repeated design). Analyses of variance were carried out using the General Linear Model module of STATISTICA 6.0 (StatSoft Inc., Tulsa, Okla., USA). Significance levels of multiple tests were adjusted after Bonferroni (Rice 1989).

The functional relationships between the dietary sterol content and the somatic (g) or population growth rates (r) were expressed as Monod curves (Monod 1950) modified with a threshold S_0 for zero growth (Rothhaupt 1988):

$$g = g_{max} \frac{c - S_0}{c - S_0 + K_S} \qquad\qquad r = r_{max} \frac{c - S_0}{c - S_0 + K_S}$$

where g_{max} and r_{max} are the maximum growth rates (d^{-1}), c is the resource concentration (μg mg C^{-1}), S_0 is the threshold concentration for zero growth (μg mg C^{-1}), and K_S is the half saturation constant (μg mg C^{-1}).

Estimations of the incipient limiting level (ILL) were based on comparisons of growth rates using 1-way ANOVAs. The sterol concentration that led to a significant decrease in growth rate with decreasing sterol supply was defined as ILL. To test statistically whether different food categories affected the mortality of *Daphnia*, a Generalized Linear Model (GLM) with the logit function as the link function for binominal distribution was used (R 2003, Version 1.8.1).

Results

Juvenile maternal growth—Increasing proportions of *Synechococcus elongatus* in the food suspensions in which ≥50% of the total organic carbon was supplied by the cyanobacteria produced a significant decline in juvenile somatic growth rates (g) of both *Daphnia* species [Tukey's HSD following ANOVA, $F_{5,12}=1490$ (*D. magna*); $F_{5,12}= 1097$ (*D. galeata*); $P<0.001$; Fig. 1A and B]. In both species, high somatic growth rates were observed when the food consisted of 100% Scen; growth on 100% Syn was generally poor. Supplementation of *Synechococcus* with cholesterol increased the somatic growth rates to levels observed in the 20% Scen + 80% Syn treatment.

Maternal dry mass—Maternal dry mass of both *D. magna* and *D. galeata* fed 100% Scen increased significantly from first through second to third reproduction (Tukey's HSD following ANOVA, Table 1) and were in general lower with decreasing sterol availability (higher percentage of Syn; Fig. 2A and B). Neither species showed an increase in maternal dry mass with successive reproduction cycles when sterols were severely limiting (20% Scen + 80% Syn). Both species responded to sterol supplementation of the otherwise sterol-free diet (100% Syn) with reproduction and maternal dry masses that were equal or even higher (in the third clutch) than those observed with animals feeding on 20% Scen + 80% Syn (Fig. 2A and B). These similar responses to growth with severely limiting sterol concentrations were contrasted by the responses observed with a less severe limitation in dietary sterols, which, in general, indicated a higher susceptibility of mothers of *D. galeata* than of *D. magna* to low sterol supply. Under moderately limiting sterol concentrations (80% Scen + 20% Syn), maternal dry mass increased with successive reproduction cycles in *D. magna,* but not in *D. galeata* in the second to third reproduction cycle (Fig. 2A and B). During growth with intermediate sterol concentrations (50% Scen + 50% Syn), the dry mass of *D. magna* was reduced at the third reproduction cycle, whereas the dry mass of *D. galeata* was already reduced at the first reproduction cycle in comparison to dry masses achieved on less limiting growth conditions (Tukey's HSD following ANOVA, Table 1).

Offspring dry mass—The dry mass of individual neonates of *D. magna* (Fig. 2C) was higher in the second clutch than in the first clutch under all growth conditions (Tukey's HSD following ANOVA, Table 1). The dry mass of individuals was higher in the third clutch than in the second clutch when the maternal diet contained 80% and 50% Scen. Between treatments, no differences in the individual dry mass of first, second, and third clutch neonates were found, except for the neonates fed on cholesterol-supplemented *Synechococcus*, which were significantly lighter in all clutches. The dry mass of second-clutch *D. galeata* neonates was higher than the dry mass of first-clutch neonates under all growth conditions (Fig. 2D), but the dry mass did not increase further in third-clutch individuals (Tukey's HSD following ANOVA, Table 1). First-, second-, and third-clutch neonates of *D. galeata* whose mothers fed on cholesterol-supplemented *Synechococcus* were also significantly lighter in all clutches.

Number of offspring—*D. magna* and *D. galeata* showed similar dietary effects on numbers of offspring during growth in the absence of sterol limitation (highest numbers of offspring on 100% Scen) and under severe sterol limitation (lowest numbers of offspring on 20% Scen + 80% Syn; Fig. 2E and F). Neither *D. galeata* nor *D. magna* reproduced on a diet of 100% Syn. Reproduction on cholesterol-supplemented *Synechococcus* was similar to that during growth with severe sterol limitation (20% Scen + 80% Syn; Fig. 2E and F), in which no increase in the number of offspring with successive reproduction cycles was observed. With moderate sterol reduction (80% Scen + 20% Syn), the number of *D. magna* offspring increased with successive reproduction cycles, but numbers of *D. galeata* did not increase from the second to third reproduction cycle (Tukey's HSD following ANOVA, Table 1). With intermediate sterol reduction (50% Scen + 50% Syn), the number of *D. magna* offspring was higher in the second clutch than in the first clutch, whereas there was no significant difference in the number of *D. galeata* offspring (Fig. 2E and F).

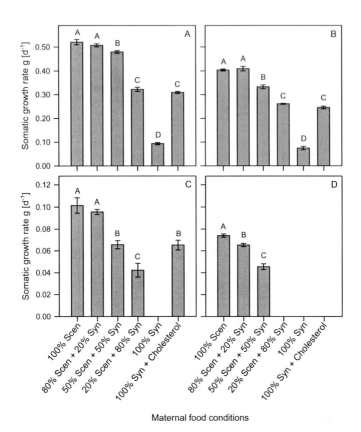

Fig. 1 Juvenile somatic growth of *Daphnia magna* (A) and *Daphnia galeata* (B) mothers reared on 100% *Scenedesmus obliquus* (100% Scen), on increasing proportions of *Synechococcus elongatus* (Syn) mixed with Scen, and on *Synechococcus elongatus* supplemented with cholesterol (100% Syn + Cholesterol). Juvenile somatic growth of the first clutch offspring of mothers grown on the corresponding food regimes was determined by feeding a 100% *S. elongatus* diet (C, *D. magna*; D, *D. galeata*). Daphnids fed on 100% *S. elongatus* did not reproduce. *D. galeata* neonates whose mothers had fed on 20% Scen + 80% Syn or on *S. elongatus* supplemented with cholesterol did not survive the 6-day experiment. Data are means of three replicates per treatment; error bars indicate SD. Bars labeled with the same letters are not significantly different (Tukey's HSD following ANOVA).

Table 1. Results of analysis of variance (ANOVA, repeated measures designs with clutch as repeated factor) of maternal dry mass, offspring dry mass, and the number of offspring in the first, second, and third clutch of *Daphnia magna* and *Daphnia galeata* reared on various food regimes. Raw data met assumptions for ANOVA (*** indicate significance after Bonferroni adjustment $P<0.0083$).

	D. magna				*D. galeata*			
	SS	df	F	P	SS	df	F	P
Maternal dry mass								
Food regime	1283464	4	336.44	***	6511.22	4	990.31	***
Error	9537	10			16.44	10		
Clutch	146309	2	313.90	***	762.79	2	184.65	***
Clutch×Food regime	63546	8	34.08	***	385.92	8	23.36	***
Error	4661	20			41.31	20		
Offspring dry mass								
Food regime	24.69	4	71.79	***	2.57	4	75.56	***
Error	0.86	10			0.09	10		
Clutch	30.59	2	283.98	***	1.42	2	105.22	***
Clutch×Food regime	3.20	8	7.42	***	0.22	8	4.09	***
Error	1.08	20			0.13	20		
Number of offspring								
Food regime	4794.49	4	253.01	***	309.48	4	1389.87	***
Error	47.38	10			0.56	10		
Clutch	2431.98	2	757.63	***	19.15	2	95.84	***
Clutch×Food regime	1793.94	8	139.72	***	39.83	8	49.83	***
Error	32.10	20			2.00	20		

Offspring growth—In order to assess the maternal effects of sterol-limited growth on offspring performance, mothers were raised on various mixtures of *Scenedesmus* and *Synechococcus* and juvenile somatic growth rates of first clutch neonates on a 100% Syn diet were determined. *D. magna* and *D. galeata* offspring exhibited the same pattern of growth as their experimental mothers (Fig. 1C and D). The growth of offspring of both species declined with decreasing sterol content of the maternal diet. However, a significant decrease in growth of *D. galeata* offspring occurred already with moderate sterol reduction (80% Scen + 20% Syn) in the maternal diet (Tukey's HSD following ANOVA, $F_{2,6}=173$; $P<0.001$), whereas the same maternal diet had no effect on the growth of *D. magna* offspring. A significant decline in *D. magna* offspring growth was observed first under intermediate sterol reduction (50% Scen + 50% Syn) in the maternal diet (Tukey's HSD following ANOVA, $F_{4,10}=71$; $P<0.001$; Fig. 1C and D).

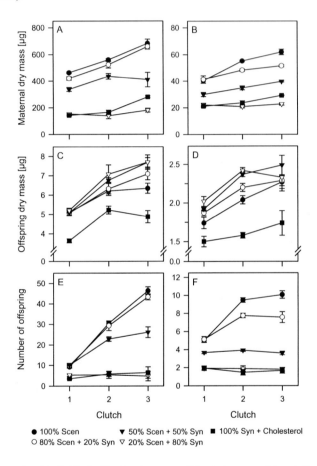

Fig. 2 Maternal and individual offspring dry mass of *Daphnia magna* (A and C) and *Daphnia galeata* (B and D) at first, second, and third reproduction grown on 100% *Scenedesmus obliquus* (100% Scen), on increasing proportions of *Synechococcus elongatus* (Syn) mixed with Scen, and on cholesterol-supplemented *S. elongatus* (100% Syn + Cholesterol). The number of viable offspring in the first, second, and third clutch is given in E (*D. magna*) and F (*D. galeata*). Daphnids fed on 100% Syn did not reproduce (not shown). Error bars indicate SD.

D. galeata offspring whose mothers had been fed with 80% Syn or with cholesterol-supplemented *Synechococcus* did not survive the 6-day experiment, whereas *D. magna* offspring performed well under these conditions, which provided further evidence for more pronounced maternal effects in *D. magna* than in *D. galeata*.

General mortality—The probability of survival, used to estimate population growth, was calculated from the mortality that occurred in the different treatments. Neither *Daphnia* species reached maturity when grown on 100% Syn and the mortality was comparatively high (Fig. 3); *D. galeata* survival was significantly lower than that of *D. magna* (GLM, $P<0.001$). The mortality of both species grown on 20% Scen + 80% Syn and on cholesterol-supplemented *Synechococcus* was significantly lower than when grown on 100% Syn (GLM, $P<0.001$). The mortality of the two species during growth on 20% Scen + 80% Syn and on cholesterol-supplemented *Synechococcus* did not differ (GLM, *D. magna*, $P=0.131$; *D. galeata*, $P=0.410$). No mortality was observed with either species on food regimes containing $\geq$50% Scen.

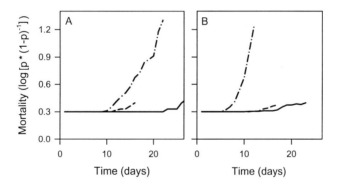

Fig. 3 Mortality of *Daphnia magna* (A) and *Daphnia galeata* (B) grown on 20% Scen + 80% Syn (dashed line), 100% Syn (dashed-dotted line), and cholesterol-supplemented *Synechococcus elongatus* (solid line) until the release of the third clutch. Animals grown on 100% Syn did not reach maturity. No mortality was observed in food regimes containing $\geq$50% Scen (data not shown). p = Probability of mortality.

Population growth—The effect of increasing proportions of cyanobacterial carbon in the diet on population growth (Fig. 4) reflected the results observed for somatic growth. Population growth rates were significantly lower when the *Synechococcus* content reached 50% (*D. magna*) or 20% (*D. galeata*) [Tukey's HSD following ANOVA, $F_{4,10}=314$ (*D. magna*), $F_{4,10}=1123$ (*D. galeata*); $P<0.001$]. The daphnids were not able to reproduce on a

100% Syn diet, and thus no positive population growth occurred. In contrast, cholesterol supplementation of *Synechococcus* led to population growth rates of 0.16 d^{-1} for *D. magna* and 0.11 d^{-1} for *D. galeata*. However, these population growth rates were significantly lower than those observed on the 20% Scen + 80% Syn diet in both species.

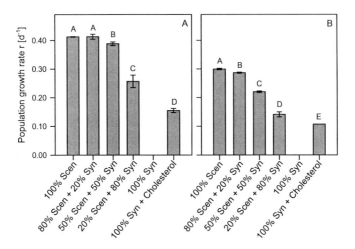

Fig. 4 Estimated population growth of *Daphnia magna* (A) and *Daphnia galeata* (B) grown on 100% *Scenedesmus obliquus* (100% Scen), on increasing proportions of *Synechococcus elongatus* (Syn) mixed with Scen, and on *S. elongatus* supplemented with cholesterol (100% Syn + Cholesterol). Data are means of three replicates per treatment; error bars indicate SD. Bars labeled with the same letters are not significantly different (Tukey's HSD following ANOVA).

Growth responses to sterol supply—The total sterol content of the food suspensions was calculated by adding up the individual amounts of the three major phytosterols detected in the green alga *Scenedesmus*. Chondrillasterol [(24E)-5α-poriferasta-7,22-dien-3β-ol, 60%], fungisterol (5α-ergost-7-en-3β-ol, 26%), and 22-dihydrochondrillasterol (5α-poriferast-7-en-3β-ol, 13%) were identified as the principal sterols in *Scenedesmus*, which is in accordance with published data (Wright et al. 1980; Cranwell et al. 1990; Rzama et al. 1994). It was assumed that these sterols can be converted to cholesterol by *Daphnia* as has been shown by Martin-Creuzburg & von Elert (2004), and therefore they were considered as cholesterol equivalents (according to von Elert et al. 2003). Thus, *Scenedesmus* contained 6.8 ± 0.6 μg (mean ± SD, n = 3) cholesterol equivalents per mg carbon. *Synechococcus* did not contribute to the total sterol content of the food suspensions since no sterols were found in the cyanobacterium. The amount of sterols found in the 20% Scen + 80% Syn food regime and in the sterol-supplemented diet did not differ. The kinetics of sterol-limited growth, expressed as a saturation curve, was applied to a modified Monod model. The somatic growth rates and the estimated population growth rates of both *Daphnia* species were highly correlated to the dietary sterol content (Fig. 5). Albeit not considered in the Monod model, somatic growth rates, and to a lesser extent population growth rates, on the sterol-supplemented diet fitted well to the correlation of dietary sterol content vs. growth rate. If the data from the sterol supplementation treatment were included in the model, the model parameters would lie within the 95% confidence interval of the values calculated without the supplemented treatment. This further corroborates the correlative evidence for growth limitation by sterols. Differences in the extent with which somatic and population growth rates on the sterol-supplemented diet fitted to the correlation of dietary sterol content vs. growth rate suggest that reproduction is less sensitive to sterol limitation and that other factors (e.g. PUFAs) may become limiting in later life stages.

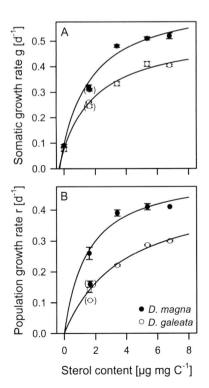

Fig. 5 Somatic (A) and population growth (B) of *Daphnia magna* and *Daphnia galeata* in response to the dietary sterol content. The regression line was calculated using a modified Monod model (Nonlinear estimations after Gauss-Newton: somatic growth $F_{(3,14)}=31.00$ *D. magna*; $F_{(3,14)}=31.52$ *D. galeata*; population growth $F_{(3,14)}=20.72$ *D. magna*; $F_{(3,14)}=18.47$ *D. galeata*; $P<0.001$). The growth of animals reared on sterol-supplemented *Synechococcus elongatus* was not considered in the Monod model; symbols are given in parentheses. Error bars indicate SD.

In both species, a significant decline in the somatic growth rates with decreasing sterol supply was observed for sterol levels <5.4 µg per mg of dietary carbon (Tukey's HSD following ANOVAs). Population growth rates of *D. magna* were reduced at sterol levels <3.4 µg per mg of dietary carbon, whereas population growth rates of *D. galeata* were reduced at sterol levels <6.8 µg per mg of dietary carbon (Tukey's HSD following ANOVAs).

Discussion

The trophic structure of aquatic food webs can be determined by the availability of essential nutrients. Recently, the absence of sterols has been considered as one potentially important factor that is able to constrain the carbon transfer efficiency between cyanobacteria as primary producers and herbivore consumers (von Elert et al. 2003). In the present study, we confronted two *Daphnia* species (*D. magna* and *D. galeata*) with increasing proportions of dietary cyanobacterial carbon in order to assess the relevance of dietary sterols for life history traits of *Daphnia*.

The somatic growth rates (g) of both *Daphnia* species were significantly lower during growth on increasing proportions of *Synechococcus* in the food, which indicated the inadequacy of cyanobacterial carbon to support *Daphnia* growth. A higher proportion of *Synechococcus* in the food is equivalent to a lower amount of dietary sterols since no sterols were found in the cyanobacterium. On a diet of 100% Syn, somatic growth of both *Daphnia* species was poor. However, supplementation of the cyanobacterium with cholesterol enhanced the somatic growth rates, which indicated that the animals were limited by the availability of sterols. Although supplementary cholesterol significantly improved the food quality of *Synechococcus*, the resultant somatic growth rates were still lower than on 100% Scen. These differences might be the consequence of an insufficient sterol supply, which is favoured by the finding that somatic growth on sterol-supplemented *Synechococcus* fitted well to the correlation of dietary sterol content vs. growth rate. We could not entirely exclude that, to some extent, PUFA limitation also constrained *Daphnia* growth, since both sterol and PUFA content of the food mixtures decreased with increasing amounts of cyanobacterial carbon. However, previous work has shown that daphnids are primarily sterol limited when grown on *Synechococcus* and that growth of *Daphnia* is enhanced by PUFAs only when the shortage of sterols has been overcome by sterol supplementation (von Elert et al. 2003). This is corroborated by the supplementation treatment which clearly shows sterol limitation as the major constraint. Von Elert et al. (2003) estimated that *Daphnia galeata* requires 20% *Scenedesmus* in the diet to compensate for the low sterol content of a cyanobacterial food.

Here we showed that sterol limitation occurs even on a diet containing 50% of the green alga. This discrepancy in the observed sensitivity to sterol limitation might be due to the lower sterol content of the green algae in the present study (6.8 vs. 10 µg per mg C; von Elert et al. 2003).

Examination of the factors that limit somatic growth is useful to ascertain individual physiological demands, but is potentially insufficient to predict population dynamics. Therefore, we determined life history parameters of *Daphnia* from birth through the third reproductive cycle to estimate the role of dietary sterols in population growth.

Size at maturity is a key parameter often considered in studies dealing with variations of life history traits and their effects on fitness. In *Daphnia*, size at maturity is determined by the size at birth, the number of juvenile instars (e.g., Tessier and Consolatti 1989), and environmental factors (e.g., Porter et al. 1983, Pace et al. 1984). Deteriorating environmental conditions reduce the size at maturity (e.g., Lynch 1989). In the present study, the dry mass (which is correlated with body size) at maturity of both *Daphnia* species was lower with higher proportions of dietary cyanobacterial carbon. Relative to the dry mass of animals fed a 100% Scen diet, this effect increased with the reproductive state and was most pronounced at the third reproduction cycle. The dry mass of animals reared on a 80% Syn diet did not increase in successive reproduction cycles. This might be caused by differing metabolic demands for growth and survival. Maintenance metabolism is based primarily on the uptake of energy (in terms of carbon), while somatic growth requires energy and a variety of essential nutrients (Sterner and Robinson 1994). The food quantity provided in the present study was well above the incipient limiting level, and *Synechococcus* is well-assimilated by *Daphnia* (Lampert 1977*a,b*), which makes energy limitation rather unlikely. The lower amount of sterols in the 80% Syn diet is apparently not sufficient to support body growth; instead, sterols might be allocated to reproduction or used to replace degraded structural material, such as cell membranes.

While effects of food quantity on somatic growth and reproduction are highly correlated in juvenile *Daphnia* (Lampert and Trubetskova 1996), effects of food quality on growth and reproduction can differ (Urabe and Sterner 2001; Becker and Boersma 2003). In the present study, the number of neonates decreased with increasing proportions of dietary *Synechococcus*. This might be partly due to the decreasing size of the experimental mothers since clutch sizes are known to be size-dependent (e.g., Lampert 1993), but it might also be a consequence of the reduced availability of sterols, which are not only needed for somatic growth, but also for reproduction (Martin-Creuzburg and von Elert 2004). In contrast to the lower numbers of offspring, the dry weights of the neonates were slightly, but not significantly, enhanced with increasing proportions of *Synechococcus* in the maternal food. This might be considered as an adaptive response to lower food quality that gives *Daphnia* offspring an advantage to survive poor food conditions. Gliwicz and Guisande (1992) have demonstrated that *Daphnia* mothers are capable of assessing food levels and accordingly adjust their fractional per-offspring allocation of reproductive resources. They have shown that high food levels lead to the production of large clutches with smaller eggs and low food levels lead to the production of small clutches with larger eggs that are more resistant to starvation. In the present study, where food quantity was constant throughout experiments, such a trade-off between clutch size and body mass of the hatched neonates (which is strongly related to egg size, Guisande and Gliwicz 1992) is conceivably a response to the decreasing dietary sterol content.

Although size at birth is an important fitness determinant (Tessier and Consolatti 1989; Lampert 1993), maternal effects on subsequent life stages should be considered when offspring performance and population dynamics are analyzed since offspring quality is not necessarily reflected in size at birth (Sakwinska 2004). Therefore, we reared first-clutch neonates of experimental mothers grown with the various food regimes on a 100% Syn diet. In this way, the pattern of growth of the offspring was, at least partly, a reflection of that of their mothers on an overall lower level. Somatic growth rates of the neonates were lower with increasing proportions of *Synechococcus* in the maternal food. This is in accordance with previous data, which showed that maternal consumption of a high quality food (*Rhodomonas*) increased neonate fitness relative to a low quality maternal food (*Microcystis*) when neonates were reared on *Microcystis* (Brett 1993). The data presented here imply a limited ability of daphnids to adjust the amount of sterols allocated to reproduction since low sterol levels in the maternal food should lead to increased sterol allocation to offspring tissues to enhance offspring survival. Instead, the results rather suggested a per-offspring allocation of sterols proportional to the availability of dietary sterols. As a consequence, insufficient amounts of sterols resulted in a limitation of both maternal and offspring growth (on a sterol-free diet). *D. galeata* neonates whose mothers had fed on 80% Syn or on sterol-supplemented *Synechococcus* did not survive the experimental period (6 days), which showed that limited amounts of sterols, adequate for low maternal growth and reproduction, are not sufficient to support offspring survival. In contrast, no mortality of *D. magna* neonates occurred within the 6-day experiment, which pointed to differences in sterol requirements between the two species. This was corroborated by the finding that survival of *D. galeata* mothers on a 100%

Syn diet was significantly lower than that of *D. magna*, which seems to be less sensitive to sterol limitation. These differences in the sensitivity to sterol limitation might be owing to differences in body size and in the storage capacity for sterols.

Effects of the diminishing availability of sterols on somatic and population growth followed more or less the same pattern. Differences in the growth response to sterol supplementation between somatic (g) and population growth rates (r) were found. Somatic growth on sterol-supplemented *Synechococcus* was as good as growth on a 20% Scen diet, whereas the estimated population growth on sterol-supplemented *Synechococcus* was significantly lower than on a 20% Scen diet. This is mainly caused by the prolonged time needed in the sterol-supplemented food regime to reach the third reproductive state (a parameter considered in the Euler-Lotka equation), which might indicate that other factors become limiting in subsequent life stages. However, it should be recognized that sterol supplementation of *Synechococcus* led to a positive population growth rate, whereas animals grown on unsupplemented *Synechococcus* did not even reproduce.

The analysis of the functional response of sterol-limited growth revealed a high correlation between the sterol content of the diet and the somatic and population growth rates. Growth rates (g and r) achieved on the cholesterol-supplemented *Synechococcus* diet were also highly correlated to the amount of sterol found in the supplemented cyanobacterium, which emphasizes the role of sterols as the principal limiting factor in this system. In the field, the performance of *Daphnia* populations is subject to multiple factors, including food availability and predation. However, the dietary sterol content should be considered when analyzing *Daphnia* growth in natural environments. In the past, the food thresholds concept (Lampert 1977c) was often used to predict *Daphnia* growth on limiting resources. Two types of threshold then have to be distinguished: The individual threshold (S_0 for g) is the amount of food at which assimilation equals respiration, resulting in a zero mass change of an individual. The population threshold (S_0 for r) is the amount of food at which reproduction compensates mortality, resulting in a zero change of population size. This concept has also been used to describe qualitative aspects of food limitation (Gliwicz and Lampert 1990, Sterner and Robinson 1994). In the present study, values of S_0 for juvenile somatic growth were negative, since positive growth rates were observed even on a sterol-free diet. This indicates sterol allocation to offspring and a limited storage capacity of sterols, but it does not allow the determination of an individual threshold for sterol-limited body growth. In contrast, the sterol thresholds for population growth were close to zero in both species. Although unlikely to occur in the field, such low sterol levels would not allow positive population growth and should finally lead to the extinction of the population.

Furthermore, we estimated the incipient limiting level (ILL) at which maximum growth passes into sterol-limited growth — a value potentially important to predict the fate of *Daphnia* populations in the field. The presented data suggest that, in both species, a sterol content of <5.4 µg per mg of dietary carbon results in reduced somatic growth. At the population level, population growth rates of *D. magna* will be lower at sterol contents of <3.4 µg per mg of dietary carbon, whereas population growth rates of *D. galeata* will be lower already at sterol contents of <6.8 µg per mg of dietary carbon (or even higher since no maximum growth rate plateau was reached). Differences between *D. magna* and *D. galeata* in their ability to grow on *Synechococcus* have previously been reported (DeMott and Müller-Navarra 1997; DeMott 1998). The authors suggest a reduced ability of *D. galeata* to utilize a PUFA-deficient diet. Data provided here point to interspecific differences in sterol requirements between *D. magna* and *D. galeata*. In *D. magna*, a lower incremental increase in dietary sterol content is required for the same increase in growth as in *D. galeata* (slope of regression line, Fig. 5), which suggested a lower sterol requirement of *D. magna*. The ability to cope with the low availability of sterols might be an adaptation of *D. magna* for living in eutrophic ponds (Lampert 1991), which are often dominated by cyanobacterial assemblages. Furthermore, the data presented suggested that interspecific differences in sterol requirements potentially affect the outcome of competition between coexisting *Daphnia* species when sterols are in short supply. To verify the suitability of the dietary sterol content to predict *Daphnia* growth in natural environments, further studies are needed to link growth responses of *Daphnia* to seston characteristics (e.g., sterol content).

The concept presented implies that populations decline with decreasing availability of sterols and vice versa, a process that might occur during phytoplankton species succession. In particular, it might explain the often-observed absence of *Daphnia* species during cyanobacterial blooms. In the field, effects of sterol limitation on population growth are probably even more evident since maternal effects on later life stages are not considered in the Euler-Lotka equation used here to estimate effects on population growth. In the present study, we documented that decreasing amounts of dietary sterols in the maternal food reduce the potential of neonates to grow under poor food conditions.

In summary, we showed that the dietary sterol content unambiguously affects life history traits of *Daphnia*. Somatic and population growth rates, maternal dry weights, the number of viable offspring, and the probability of survival were significantly reduced with the diminishing availability of sterols. This suggested a strong impact of dietary sterols on population dynamics of the herbivorous grazer. Sterol-limited growth might lead to a reduced energy transfer at the plant–animal interface and therefore to a decoupling of primary and secondary production. We are aware that herbivores like *Daphnia* face a complex pattern of nutritional challenges in natural environments, but sterol limitation might be one important factor with the potential to determine the structure of aquatic food webs.

Acknowledgments—This study was supported by the German Research Foundation (DFG, El 179/4-2). We thank S. Schumacher for experimental assistance, K.A. Brune for editing the English, and two anonymous reviewers for valuable comments on an earlier draft of this manuscript.

Chapter 3

Impact of 10 dietary sterols on growth and reproduction of *Daphnia galeata*

Dominik Martin-Creuzburg and Eric von Elert

Abstract—In crustaceans, cholesterol is an essential nutrient, which they must directly obtain from their food or by bioconversion from other dietary sterols. Eukaryotic phytoplankton contain a great variety of sterols that differ from cholesterol in having additional substituents or different positions and/or number of double bonds in the side chain or in the sterol nucleus. In this study, we investigated to what extent these structural features affect the growth and reproduction of *Daphnia galeata* in standardized growth experiments with the cyanobacterium *Synechococcus elongatus* supplemented with single sterols as food source. The results indicated that Δ^5 (sitosterol, stigmasterol, desmosterol) and $\Delta^{5,7}$ (7-dehydrocholesterol, ergosterol) sterols meet the nutritional requirements of the daphnids, while the Δ^7 sterol lathosterol supports somatic growth and reproduction to a significantly lower extent than cholesterol. Dihydrocholesterol (Δ^0) and lanosterol (Δ^8) did not improve the growth of *D. galeata,* and growth was adversely affected by the Δ^4 sterol allocholesterol. Sterols seem to differ in their allocation to somatic growth and reproduction. Thus, structural differences of dietary sterols have pronounced effects on life-history traits of *D. galeata*.

Keywords—food quality, cyanobacteria, dietary sterols, cholesterol, *Daphnia galeata*

Introduction

The transfer of energy from primary producers to higher trophic levels is an important factor that determines the trophic structure of aquatic food webs. At the phytoplankton–zooplankton interface, the efficiency of carbon transfer is highly variable. This variation can be attributed to the changing nutritional value of phytoplankton assemblages. Nutritional inadequacy can be due to toxicity (Lampert, 1981*a,b*), digestive resistance (Porter and McDonough, 1984), or mineral (Elser et al., 2001) or biochemical composition of phytoplankton species and can result in a decoupling of primary and secondary production. The biochemical composition of phytoplankton, in particular the content of polyunsaturated fatty acids (PUFAs), has been discussed as being potentially limiting for *Daphnia* growth (Ahlgren et al., 1990; Müller-Navarra, 1995; Wacker and Von Elert, 2001).

PUFAs are of special importance for freshwater zooplankton nutrition in lakes dominated by cyanobacteria, as articulated in a correlative study by Müller-Navarra et al. (2000). Cyanobacteria in general lack long-chain PUFAs (Cobelas and Lechardo, 1988; Ahlgren et al., 1992), and the well-known low carbon-transfer efficiency at the cyanobacteria–*Daphnia* interface has been suggested to be caused by a deficiency in long-chain PUFAs (Müller-Navarra et al., 2000). Supplementation of the cyanobacterium *Synechococcus elongatus* with a PUFA-rich fish oil emulsion leads to better growth and reproduction of *Daphnia* (DeMott and Müller-Navarra, 1997) and therefore supports the correlative evidence. However, Von Elert and Wolffrom (2001), have found that the absence of a non-PUFA lipid present in eukaryotic algae constrains assimilation of cyanobacterial carbon. Fish oil contains other lipids in addition to PUFAs, such as sterols, which are also essential for growth and reproduction of crustaceans (Goad, 1981). Cyanobacteria, as prokaryotes, lack or contain only traces of sterols (Hai et al., 1996, Volkman, 2003). In a previous study, we have shown that the low carbon-transfer efficiency of cyanobacteria to *Daphnia galeata* is caused by the lack of sterols in cyanobacteria (Von Elert et al., 2003).

Like all arthropods, crustaceans are incapable of synthesizing sterols de novo and therefore must acquire these essential nutrients from their diet (Goad, 1981). Crustaceans generally have a simple sterol composition with characteristic high cholesterol levels (Teshima and Kanazawa, 1971*a*, Yasuda, 1973). Cholesterol is an indispensable structural component of cell membranes and serves as a precursor for many bioactive molecules, such as ecdysteroids, which are involved in the process of molting (Goad, 1981; Harrison, 1990). However, the herbivorous cladoceran *Daphnia*, unlike carnivorous crustaceans, cannot rely on a dietary source of cholesterol because only trace amounts are found in many phytoplankton species (Nes and McKean, 1977). Eukaryotic phytoplankton contain a great variety of plant sterols (Nes and McKean, 1977; Volkman, 2003), which can be distinguished from cholesterol by their chemical structure. These phytosterols are often characterized by additional substituents or by the position and/or number of double bonds in the side chain or in the sterol nucleus (Piironen, 2000). The crustaceans examined to date are capable of converting dietary sterols to cholesterol (Teshima, 1971; Teshima and Kanazawa, 1971*b*; Ikekawa, 1985; Harvey et al., 1987), but not all sterols are suitable precursors for the synthesis of cholesterol (Teshima et al., 1983).

Under field conditions, the diet of the nonselectively suspension-feeding *D. galeata* is complex. The diet usually consists of phytoplankton, protozoa, bacteria, and detritus in varying ratios. Depending on the composition of their diet, the cladocerans are provided with a large variety of sterols in different quantities. De Lange and Arts (1999) correlated biochemical variables of natural seston with *Daphnia* growth rates and found that the sterol content is a useful tool to predict *Daphnia* growth. However, growth of the herbivorous zooplankton might not only be limited by the total sterol content itself, but also by the absence of sterols that are suitable precursors of cholesterol. In periods when phytoplankton assemblages are dominated by species with an unsuitable sterol pattern, growth and reproduction of *Daphnia* could be constrained by the low availability of suitable sterols. The first evidence that structural differences of dietary sterols can have pronounced effects on life-history traits of arthropods has been found in terrestrial systems. Behmer and Grebenok (1998) pointed out that growth and fecundity of the moth *Plutella xylostella* was affected by dietary sterols. Further on, it was recently demonstrated that sterols with double bonds at Δ^7 and/or Δ^{22} (Figure 1) failed to support development of different grasshopper species and that survival of the grasshopper *Schistocerca americana* was constrained by the ratio of suitable to unsuitable sterols in their diet (Behmer and Elias, 2000). Consistently, the development of marine copepods was negatively affected by Δ^7 sterols, whereas Δ^5 sterols allowed a rapid development of the copepods (Klein Breteler et al., 1999). Comparable investigations on the structural requirements of freshwater zooplankton with regard to sterols are missing to date.

The aim of this study was to investigate to what extent structural features of sterols, such as the alkylation of the side chain or the presence or absence of double bonds, affect the nutritional value of single sterols for *Daphnia*. Standardized growth experiments of *D. galeata* with the cyanobacterium *S. elongatus* supplemented with single sterols as food source were conducted. *S. elongatus* is well assimilated by *Daphnia* (Lampert, 1977*a,b*) and does not contain any sterols. Thus, the cyanobacterium is a convenient source of carbon and a useful "transfer vehicle" for delivering sterols to the daphnids.

Materials and methods

Cultures and growth experiments—Laboratory growth experiments were conducted with a clone of *Daphnia galeata*, which was originally isolated from Lake Constance (Stich and Lampert, 1984). The green alga *Scenedesmus obliquus* (SAG 276-3a, Sammlung von Algenkulturen Göttingen, Germany) was grown in batch culture and harvested in the late-exponential phase. It was used as the food source for the stock culture of *D. galeata* and for the newborn experimental animals, which were cultured to the age of 48 h in a flow-through system prior to the growth experiments. *Synechococcus elongatus* (SAG 89.79) was grown in chemostats at a dilution rate of $0.25 \; d^{-1}$ according to Von Elert and Wolffrom (2001). *S. elongatus* and *S. obliquus* were grown in Cyano medium (Jüttner et al., 1983). Chemostat-grown cells were concentrated by centrifugation and resuspended in fresh medium. Carbon concentrations of the cyanobacterial suspensions were estimated from photometric light extinction (800 nm) using carbon-extinction equations. *S. elongatus* had a molar C:N:P ratio of 121:23:1. Growth experiments were carried out at 20°C in glass beakers filled with 0.5 L of filtered lake water (0.45 µm pore-sized membrane filter) containing $2 \; mg \; C \; L^{-1}$ *S. elongatus*. The 48-h-old juveniles (released from the third clutch within 10 h) were transferred from the flow-through system into these beakers. The food suspensions were renewed daily within the four days of the experimental period. Somatic growth rates (g) were determined as the increase in dry weight (W) during the experiments using the equation:

$$g = (\ln W_t - \ln W_0)/ t.$$

Subsamples of the experimental animals were taken at the beginning (W_0) and at the end (W_t) of an experiment. The subsamples consisting of ~15 juveniles were dried for 12 h and weighed on an electronic balance (Mettler UMT 2; ± 0.1 µg). Each treatment consisted of three replicates with 15 animals each, and growth rates were calculated as means for each treatment.

Supplementation of sterols—Sterols used for supplementation are given in Table 1, they were selected according to their chemical structure and their natural occurrence. To enrich *S. elongatus* with sterols, 10 mg bovine serum albumin (BSA) was dissolved in 5 ml of ultra-pure water, and 200 µl of an ethanolic stock solution of the free sterol ($2.5 \; mg \; ml^{-1}$) was added. Subsequently, 4 mg particulate organic carbon (POC) of the *S. elongatus* stock solution was added to each solution, and the volume was brought to 40 ml with Cyano medium. The resulting suspension was incubated on a rotary shaker (100 revolutions min^{-1}) for 4 h. Surplus BSA and free sterols were removed by washing the cells three times in 10 ml fresh medium according to Von Elert (2002). The resulting *S. elongatus* suspension was used as food in the growth experiments.

Table 1. Nomenclature of sterols supplemented to the *Daphnia galeata* food source, *Synechococcus elongatus*.

Trivial name	IUPAC name	Formula	Commercial source
Cholesterol	Cholest-5-en-3β-ol	$C_{27}H_{46}O$	Sigma C-8667
Stigmasterol	Stigmasta-5,22-dien-3β-ol	$C_{29}H_{48}O$	Sigma S-2424
Sitosterol	Stigmast-5-en-3β-ol	$C_{29}H_{50}O$	Sigma S-1270
Ergosterol	(22E)-Ergosta-5,7,22-trien-3β-ol	$C_{28}H_{44}O$	Sigma E-6510
Lathosterol	5α-Cholest-7-en-3β-ol	$C_{27}H_{46}O$	Sigma C-3652
Dihydrocholesterol	5α-Cholestan-3β-ol	$C_{27}H_{48}O$	Sigma D-6128
Lanosterol	5α-Lanosta-8,24-en-3β-ol	$C_{30}H_{50}O$	Sigma L-1504
Allocholesterol	Cholest-4-en-3β-ol	$C_{27}H_{46}O$	Steraloids C6100
7-Dehydrocholesterol	Cholesta-5,7-dien-3β-ol	$C_{27}H_{44}O$	Steraloids C3000
Desmosterol	Cholesta-5,24-dien-3β-ol	$C_{27}H_{44}O$	Steraloids C3150

Analyses—Sterols were analyzed from approximately 0.5 mg POC of the food suspensions filtered on precombusted GF/F filters or from 60–80 animals washed twice with ultra-pure water. Lipids were extracted three times with dichloromethane:methanol (2:1, vol/vol). After saponification with 0.2 mol L^{-1} methanolic KOH (70°C, 1 h) and addition of ultra-pure water, the neutral lipids (sterols) were partitioned into iso-hexane:diethyl ether (9:1, vol/vol). The sterols were analyzed as free sterols with a gas chromatograph (HP 6890) equipped with an HP-5 capillary column (Agilent) and a flame ionization detector. The carrier gas (helium; purity 5.0) had a flow rate of 1.5 ml min^{-1}. The temperature was raised from 150 to 280°C at 15°C min^{-1} and increased to 330°C at 2°C min^{-1}. The final temperature was held for 5 min. Sterols were quantified by comparison to 5α-cholestan, which was used as an internal standard and identified using a gas chromatograph–mass spectrometer (Finnigan MAT GCQ) equipped with a fused silica capillary column (DB-5MS, J&W). Spectra were recorded between 60 and 400 amu in the EI ionization mode. POC was determined with an NCS-2500 analyzer (Carlo Erba Instruments).

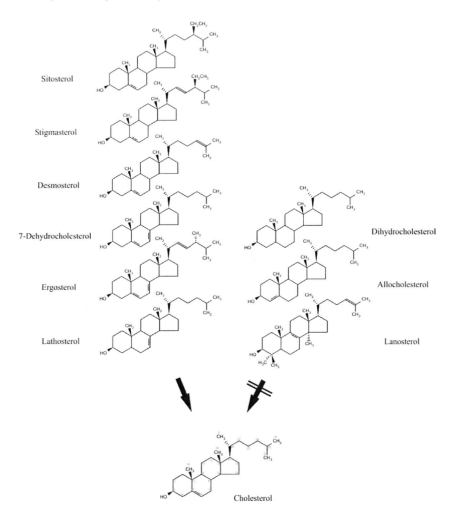

Fig. 1 Structural requirements for the conversion of dietary sterols to cholesterol in *Daphnia galeata*. Sterols on the left are suitable precursors for the synthesis of cholesterol, whereas sterols on the right are not. Potential intermediates in sterol metabolism are not shown.

Data analysis—All data were analyzed using one-way analysis of variance (ANOVA). For growth rates and clutch sizes, raw data met the assumption of homogeneity of variance; values of the supplemented sterols and cholesterol in *D. galeata* were log_{10}-transformed to meet assumptions for ANOVA. The effects of single treatments were tested by Tukey's HSD post-hoc tests. A significance level of $P = 0.05$ was applied to all statistical analyses.

Results

Growth experiments—Growth of *D. galeata* on unsupplemented *S. elongatus* was in general poor (growth rate, g = 0.07 d^{-1}). Supplementation of *S. elongatus* with sterols affected somatic growth of *D. galeata* (ANOVA, $F_{10,22}$ = 389; $P < 0.001$; Figure 2). Growth rates on cyanobacteria supplemented with stigmasterol (g = 0.30 d^{-1}), sitosterol (g = 0.32 d^{-1}), ergosterol (g = 0.32 d^{-1}), and 7-dehydrocholesterol (g = 0.30 d^{-1}) were highest and significantly different from growth rates with the other treatments (Tukey's HSD, $P < 0.05$). Supplementation with desmosterol also led to a high growth rate (g = 0.28 d^{-1}), but was significantly lower than the growth rates obtained with sitosterol, ergosterol, and 7-dehydrocholesterol.

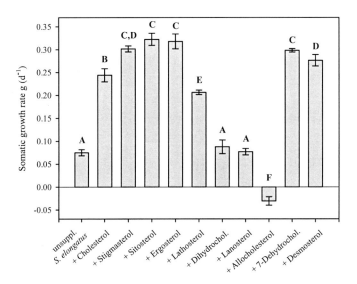

Fig. 2 Somatic growth of *Daphnia galeata* reared on *Synechococcus elongatus* unsupplemented and supplemented with single sterols. Data are means of three replicates per treatment; error bars indicate SD. Bars labeled with the same letters are not significantly different (Tukey's HSD, $P < 0.05$ following ANOVA).

Supplementation with cholesterol had a less-pronounced effect on growth (g = 0.24 d^{-1}) than supplementation with the sterols mentioned above. Dihydrocholesterol (g = 0.09 d^{-1}) and lanosterol (g = 0.08 d^{-1}) did not improve growth, compared to growth of animals reared on unsupplemented *S. elongatus* (dihydrocholesterol, $P = 0.92$; lanosterol, $P = 1$). Negative growth rates were observed after supplementation with allocholesterol ($g = -0.03$ d^{-1}).

Clutch sizes exhibited almost the same pattern as the growth rates (Figure 3). However, supplementation with 7-dehydrocholesterol led to the highest clutch size of 2.4 eggs per individual, whereas the growth rate obtained with 7-dehydrocholesterol did not differ from those obtained after supplementation with stigmasterol, sitosterol, and ergosterol. *D. galeata* fed on *S. elongatus* supplemented with stigmasterol produced 1.5 eggs per individual, which is significantly less than animals fed *S. elongatus* supplemented with sitosterol and ergosterol (Tukey's HSD, $P < 0.05$ following ANOVA, $F_{7,16} = 106$; $P < 0.001$). Although dihydrocholesterol did not improve growth, *D. galeata* did produce eggs in this treatment, with a clutch size of 0.2 eggs per individual. Animals kept on a diet supplemented with lanosterol or allocholesterol and animals fed pure *S. elongatus* did not produce eggs within the four-day experiment.

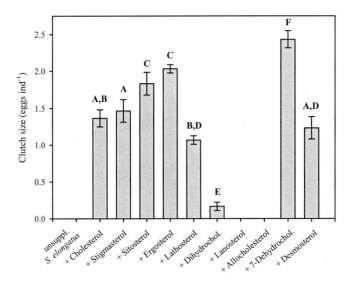

Fig. 3 Number of eggs of the first clutch of *Daphnia galeata* feeding on *Synechococcus elongatus* unsupplemented and supplemented with single sterols. Data are means of three replicates per treatment; error bars indicate SD. Bars labeled with the same letters are not significantly different (Tukey's HSD, $P < 0.05$ following ANOVA).

Sterol analysis—No sterols other than the supplemented sterols were detected in *S. elongatus*, which indicated that the supplemented sterols were not metabolically converted in the cyanobacterium. After feeding *D. galeata* four days on supplemented *S. elongatus*, all supplemented sterols could be detected in the animals (Figure 4), but the amounts per individual differed (ANOVA, F 9,20 = 45.5; P < 0.001). Cholesterol was the main sterol found in *D. galeata* in all experimental treatments. *D. galeata* fed on *S. elongatus* supplemented with cholesterol had a higher cholesterol content than animals grown on unsupplemented *S. elongatus* (Figure 4). The amounts of supplemented sitosterol, dihydrocholesterol, lanosterol, and 7-dehydrocholesterol in *D. galeata* were higher than the amounts of supplemented stigmasterol, ergosterol, and allocholesterol in the animals. Only small amounts of lathosterol and desmosterol were detected in *D. galeata* reared on food supplemented with these sterols.

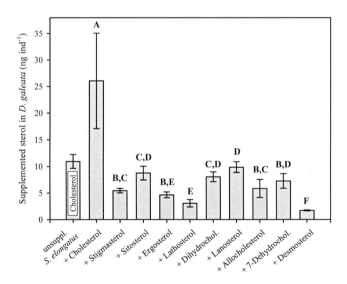

Fig. 4 Sterol content in *Daphnia galeata,* grown on *Synechococcus elongatus* unsupplemented and supplemented with single sterols. At the end of the experiment animals were analyzed for the content of the supplemented sterol. For animals grown on unsupplemented *S. elongatus,* the cholesterol content is given. Data are means of three replicates per treatment; error bars indicate SD. Bars labeled with the same letters are not significantly different (Tukey's HSD, P < 0.05 following ANOVA, F 9,20 = 45.5; P < 0.001).

Immediately prior to the experiments, newborn animals were raised for two days on the green alga *Scenedesmus obliquus*. In addition to cholesterol, small amounts of the three major phytosterols of *S. obliquus* (Von Elert et al., 2003) were detected in these animals. Although no cholesterol was found in *S. obliquus*, the cholesterol content of *D. galeata* increased after growth for an additional four days on the green alga (Table 2), which indicated that the phytosterols present in *S. obliquus* were converted to cholesterol. In contrast, the cholesterol content of *D. galeata* decreased when the two day-old animals were fed an additional four days on the unsupplemented cyanobacterium *S. elongatus* (Table 2).

Table 2. Cholesterol content of *Daphnia galeata* at the age of 2 and 6 days. The animals either were reared continuously on *Scenedesmus obliquus* or were fed with *Synechococcus elongatus* after the second day. All cholesterol contents were significantly different (Tukey's HSD following ANOVA $F_{2,6}$ = 132; $p < 0.001$). Means values of three replicates per treatment are given.

Food regime	Cholesterol content (ng ind^{-1}) ± SD
2 days on *Scenedesmus obliquus*	22.55 ± 0.57
6 days on *Scenedesmus obliquus*	54.52 ± 9.44
2 days on *Scenedesmus obliquus* / 4 days on *Synechococcus elongatus*	10.96 ± 1.30

With the assumption that cholesterol in *D. galeata* arises from the conversion of dietary sterols, the sterol-free cyanobacterium *S. elongatus* was supplemented with single sterols and the effect of the supplemented sterols on the cholesterol content of *D. galeata* was examined (Figure 5). Animals fed *S. elongatus* supplemented with ergosterol had a ten-fold higher cholesterol content (109 ng ind^{-1}) than animals grown on unsupplemented *S. elongatus* (11 ng ind^{-1}). Supplementation of cyanobacteria with ergosterol or stigmasterol led to a higher content of cholesterol in the daphnids than supplementation with cholesterol itself. Supplementation of the cyanobacterial food with sitosterol, lathosterol, dihydrocholesterol, 7-dehydrocholesterol, or desmosterol also increased the cholesterol content of *D. galeata*, which indicated that these sterols were also converted to cholesterol. Supplementation with lanosterol and allocholesterol, on the other hand, did not affect the cholesterol content of the daphnids, which suggested that neither of these sterols could be used as a cholesterol precursor by the animals.

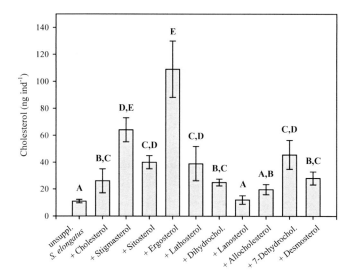

Fig. 5 Cholesterol content of *Daphnia galeata* reared on *Synechococcus elongatus* unsupplemented and supplemented with single sterols. Data are means of three replicates per treatment; error bars indicate SD. Bars labeled with the same letters are not significantly different (Tukey's HSD, $P < 0.05$ following ANOVA, $F_{10,22} = 30.6$; $P < 0.001$).

Discussion

The crustaceans examined to date are incapable of synthesizing sterols de novo — they require a dietary source of sterols to meet their basic physiological demands. In a recently published study, we have shown that poor somatic growth of *Daphnia galeata* on *Synechococcus elongatus* is due to the lack of sterols in the cyanobacterium (Von Elert et al., 2003). Supplementation of *S. elongatus* with cholesterol improved the growth of the animals, which indicates that growth of *D. galeata* was limited by cholesterol. Since herbivorous crustaceans do not find sufficient amounts of cholesterol in their diet, they need to assimilate available dietary sterols and convert them to cholesterol (Ikekawa, 1985). Eukaryotic phytoplankton usually contain a variety of sterols that can be distinguished from cholesterol by their chemical structure. These sterols are often characterized by additional substituents or by the position and/or number of double bonds in the side chain or in the sterol nucleus (Piironen, 2000).

All supplemented sterols were detected in *D. galeata,* which indicates that they were assimilated by the animals. Although single sterols were found in relatively small amounts in *D. galeata,* the amounts were too high to be exclusively derived from ingested *S. elongatus* in the gut of the animals. In cases in which only small amounts of a single supplemented sterol were found in *D. galeata,* an increased cholesterol content of the animals was observed. In contrast, when a supplemented sterol was found in higher amounts in the animals, the cholesterol content was not affected. These two patterns provide evidence for which of the supplemented sterols can be converted to cholesterol by *D. galeata*.

In animals that are capable of synthesizing cholesterol de novo, the cyclization of squalene leads to lanosterol. Lanosterol differs from cholesterol by having additional C-4 dimethyl and C-14 methyl substituents and by the location of the double bond (Δ^8) in the sterol nucleus (Figure 1). Supplementation of cyanobacteria with lanosterol did not affect growth rates and clutch sizes of *D. galeata*. Furthermore, no increase in the cholesterol content of the animals was observed. The biochemical conversion of lanosterol to cholesterol involves the loss of the methyl groups, the removal of the Δ^8 double bond, and the introduction of a double bond at Δ^5. The above findings demonstrate that *D. galeata* lacks the enzymatic ability to convert Δ^8 sterols to cholesterol. Notwithstanding our findings, the conversion of Δ^8 sterols to cholesterol was hypothesised by Harvey et al. (1987) in the marine copepod *Calanus*. This suggests taxon specific differences in the structural requirements of dietary sterols for crustaceans.

The phytosterols sitosterol and stigmasterol differ from cholesterol in having an ethyl group at C-24, and stigmasterol has an additional double bond at Δ^{22} in the side chain (Figure 1). Sitosterol and stigmasterol are commonly found in higher plants and are also present in a number of microalgae (Volkman, 2003). The synthesis of cholesterol from these sterols requires a dealkylation at C-24. An efficient phytosterol C-24-dealkylating system is found in various crustacea (Ikekawa, 1985). Teshima (1971) has described the bioconversion of sitosterol to cholesterol in the prawn *Penaeus japonicus* using ^{14}C-labeled sitosterol. Our findings that food supplemented with sitosterol or stigmasterol led to an increased cholesterol content of the animals indicates that a 24-dealkylation also occurs in *D. galeata*. Furthermore, *D. galeata* seems to be capable of saturating the additional Δ^{22} bond of stigmasterol during its transformation to cholesterol. However, sitosterol and stigmasterol improved growth more efficiently than cholesterol, which might indicate that *D. galeata* is also able to use these sterols directly without the circuitous synthesis of cholesterol and that these sterols play a yet unknown role in the metabolism of *D. galeata*.

Supplementation of cyanobacteria with desmosterol stimulated growth and egg production of *D. galeata* and increased the cholesterol content of the animals, which demonstrates that the ability to transform desmosterol to cholesterol is also present in *D. galeata*. The $\Delta^{5,24}$ diene desmosterol (Figure 1) is the terminal intermediate in the conversion of plant sterols (e.g., sitosterol and stigmasterol) to cholesterol in insects (Svoboda and Thompson, 1985). A Δ^{24} sterol reductase that reduces the double bond in the side chain, thereby converting desmosterol to cholesterol, has been found in the tobacco hornworm, *Manduca sexta* (Svoboda and Thompson, 1985). Experiments with labeled sterols have shown that the prawn *Penaeus japonicus* also possesses the ability to use desmosterol as a precursor for the synthesis of cholesterol (Teshima and Kanazawa, 1973).

D. galeata is able to convert $\Delta^{5,7}$ sterols to Δ^5 sterols, as evidenced by the large increase in the cholesterol content of the animals after supplementation of the food with 7-dehydrocholesterol ($\Delta^{5,7}$). 7-dehyrocholesterol is found in the hemolymph and in particular in Y-organs of crustaceans, where molting hormones are synthesized (Lachaise et al., 1989; Rudolph et al., 1992). In many insects, 7-dehydrocholesterol is an intermediate in the transformation of cholesterol to ecdysteroids (Rees, 1985). Several studies suggest that 7-dehydrocholesterol is formed irreversibly from cholesterol in isolated prothoracic glands (Grieneisen, 1994). Here we showed that a transformation of 7-dehydrocholesterol to cholesterol occurs in the cladoceran *D. galeata*. Assuming that cholesterol is the key sterol in crustaceans, it is surprising that 7-dehydrocholesterol improved the growth of the daphnids more efficiently than cholesterol. Synthesis of ecdysteroids from cholesterol requires the introduction of a Δ^7 bond into the sterol nucleus, which is not necessary in the direct conversion of 7-dehydrocholesterol to ecdysteroids. The conversion of labeled 7-dehydrocholesterol to labeled ecdysteroids has been demonstrated by injection experiments with various insect species (see Grieneisen, 1994) and by incubation of fractionated Y-organs of the crab *Menippe mercenaria* with the sterol (Rudolph and Spaziani, 1992). Presumably, *D. galeata* is also capable of utilizing 7-dehydrocholesterol as a direct precursor of ecdysteroids. Increased clutch sizes relative to the growth rates showed that 7-dehydrocholesterol effectively supported egg production. Although it is generally assumed that effects of food quantity on somatic growth and on reproduction are highly correlated in juvenile *Daphnia* (Lampert and Trubetskova, 1996), it has been suggested that limitation by food quality might affect somatic growth and reproduction differently. This has been shown for mineral (Urabe and Sterner, 2001) and biochemical (Becker and Boersma, 2003) aspects of food limitation. In accordance with these findings, sterols seem to differ in their allocation to somatic growth or reproduction. Further detailed investigations of sterol effects on life history are needed to reveal how these differences in allocation lead to differences in effects on fitness.

Supplementation of cyanobacteria with ergosterol resulted in a tenfold higher cholesterol content of *D. galeata* than in animals fed unsupplemented food. *D. galeata* is therefore capable of converting dietary ergosterol to cholesterol. Ergosterol, a $\Delta^{5,7,22}$ sterol, is found in most fungi, yeast, and in some species of green algae (Nes and McKean, 1977; Akihisa, 1992; Petkov and Kim, 1999). Ergosterol differs from 7-dehydrocholesterol in having an additional double bond at Δ^{22} in the side chain (Figure 1). Growth rates on food supplemented with ergosterol were as high as the growth rates reached with 7-dehydrocholesterol, sitosterol, and stigmasterol. The conversion of ergosterol to cholesterol requires the saturation at Δ^{7} in the sterol nucleus, as described for 7-dehydrocholesterol, as well as saturation at Δ^{22} in the side chain, as described for stigmasterol. Only small amounts of ergosterol were detected in animals reared on ergosterol-supplemented food, which points to high metabolic transformation rates. Teshima and Kanazawa (1971*b*) have described the bioconversion of ergosterol to cholesterol in *Artemia salina* fed on ^{14}C-labeled *Euglena gracilis*. The ability to saturate the Δ^{5} bond of a $\Delta^{5,7}$ diene, as discussed for 7-dehydrocholesterol, might also enable the direct conversion of ergosterol to ecdysteroids.

Supplementation with dihydrocholesterol, a completely saturated molecule (Δ^{0}), did not affect somatic growth of *D. galeata,* as compared with unsupplemented cyanobacteria, which indicates that a double bond in ring B is required for the conversion of dietary sterols to cholesterol (Figure 1). In contrast to somatic growth, egg production of the daphnids was positively affected by supplementation with dihydrocholesterol, which indicated the potential significance of sterols for reproduction. We are aware of only one example of the oxidation of a Δ^{0} sterol to a Δ^{5} sterol in arthropods: the firebrat, *Thermobia domestica*, is capable of synthesizing cholesterol from dihydrocholesterol (Svoboda and Thompson, 1985). Harvey et al. (1987) documented that ring-saturated stanols are poorly assimilated and that they pass unaltered through the gut of the marine copepod *Calanus*. In this study, we found significant amounts of the supplemented dihydrocholesterol in daphnid tissues, which indicates the assimilation of this stanol.

Allocholesterol and lathosterol differ from cholesterol in the position of the double bond in the sterol nucleus (Figure 1). Somatic growth of *D. galeata* was negatively affected by the supplementation with allocholesterol. A relocation of a double bond from Δ^4 to Δ^5, as required for the conversion of allocholesterol to cholesterol, seems improbable; however, we cannot exclude that a toxic effect of allocholesterol masked the enzymatic abilities of the animals. Supplementation with lathosterol increased the cholesterol content of the animals, which indicated that lathosterol was converted to cholesterol. The conversion of lathosterol to cholesterol requires a shift of a double bond from Δ^7 to Δ^5, possibly via a $\Delta^{5,7}$ intermediate, as work with mammals has shown (Nes and McKean, 1977). Prahl et al. (1984) found that, compared with Δ^5 and $\Delta^{5,7}$ sterols, Δ^7 sterols were not readily removed during passage through the gut of the copepod *Calanus*. They speculated that dietary Δ^7 sterols can be used as precursors of ecdysteroids and that the poor assimilation of these sterols provides a mechanism to avoid a haphazard production of molting hormones. Alternatively Prahl et al. (1984) suggested that *Calanus* simply lacks the ability to convert Δ^7 to $\Delta^{5,7}$ sterols and therefore the Δ^7 components are only poorly assimilated. The results of this study indicate that the Δ^7 sterol lathosterol was assimilated by *D. galeata* and converted to cholesterol. However, the observed growth rates were lower than those reached with food supplemented with cholesterol. The step Δ^7 to $\Delta^{5,7}$ involves the introduction of a double bond at Δ^5, which might be costly in terms of energy and therefore might be responsible for the lower growth rates as compared to those reached with supplementation with cholesterol.

Although this study shows that certain dietary sterols improve the somatic growth of *D. galeata*, there must be other factors that become limiting for the growth and reproduction of the herbivore, when the animals are released from sterol limitation. The maximal growth rates on sterol-supplemented *Synechococcus* ($g = 0.32$ d^{-1}) were below the almost maximal possible growth rate ($g = 0.5$ d^{-1}) of *D. galeata* fed on the green alga *Scenedesmus obliquus* (Wacker and Von Elert, 2001). Von Elert et al. (2003) already showed that the growth on cholesterol-supplemented *Synechococcus* was further improved by additional supplementation with PUFAs. Beside sterols and PUFAs there might be additional factors that determine the nutritional value of this coccal cyanobacterium to a lower extent.

Results derived from laboratory experiments are indispensable for determining the requirements of zooplankton species for single biochemical compounds, such as sterols, and provide a first step toward assessing the ecological relevance of these compounds under field conditions. During cyanobacterial blooms, the sterol content of the food will be low since only traces of sterols are found in the prokaryotes (Hai et al., 1996; Volkman, 2003). This is corroborated by the observation that the total lipid levels (with sterols as a dominant lipid class) of *Daphnia pulex* from a hypereutrophic lake are at their lowest concentration during the height of the yearly *Aphanizomenon flos-aquae* bloom (Arts et al., 1992). In a previous laboratory study, we have shown that the absence of sterols constrains the carbon transfer between cyanobacteria and *D. galeata* (Von Elert et al., 2003). Compared to cyanobacteria, which do not provide sterols in sufficient amounts, eukaryotic phytoplankton contain a large variety of sterols (Nes and McKean, 1977; Volkman, 2003). However, specific phytoplankton classes or even single species could still be deficient in sterols suitable for supporting zooplankton growth. If such species dominate the phytoplankton, sterol limitation of growth of *Daphnia* is possible. Thus, high levels of unsuitable sterols could adversely affect growth and reproduction of *Daphnia*, and can therefore be responsible for reduced fecundity and, projected at the population level, for reduced population growth.

In the field, sterols of phytoplankton can be subjected to transformation prior to their ingestion by the herbivorous crustaceans. Klein Breteler et al. (1999) have suggested that the poor quality of the chlorophycean *Dunaliella* for the development of marine copepods is due to a sterol deficiency of the alga. Furthermore, they have demonstrated that the chlorophycean food is biochemically upgraded by the heterotrophic dinoflagellate *Oxyrrhis marina* to high-quality copepod food. This trophic upgrading of food quality by an intermediary protozoan is attributed to sterol production in the dinoflagellate. The Δ^7 sterols present in *Dunaliella* do not support development of the copepods, whereas a rapid development of the copepods to the adult stage is observed when fed on *Oxyrrhis marina*, which contains primarily Δ^5 sterols. This example shows that unsuitable sterols in eukaryotic algae can constrain the development of herbivorous crustaceans. Intermediary grazers, such as protozoa, might biochemically upgrade such unsuitable phytoplankton species by adding more suitable sterols to the dietary carbon and thus determining the transfer efficiency of carbon from the microbial loop to metazoan grazers in natural systems.

In summary, this study provides evidence that sterols are essential dietary compounds that significantly affect growth and reproduction of *D. galeata*. Furthermore, the results showed that *D. galeata* is capable of converting dietary sterols to cholesterol, depending on their chemical structure. Particularly, Δ^5 and $\Delta^{5,7}$ sterols met the nutritional requirements of the animals, while the Δ^7 sterol lathosterol supported growth to a significantly lower extent than cholesterol. Dihydrocholesterol (Δ^0) and lanosterol (Δ^8) did not improve the growth of *D. galeata,* and growth was adversely affected by the Δ^4 sterol allocholesterol. Hence, structural features, particularly the configuration of the sterol nucleus, determine the nutritional value of dietary sterols. In insects, the pattern of sterol metabolism is by no means ubiquitous, and the nutritional dependency on specific sterols described for *D. galeata* might not be valid for crustaceans in general. In order to assess the ecological significance of certain sterols as potentially limiting biochemical resources, further detailed studies are required to reveal pathways and potential intermediates of sterol synthesis with regard to the nutritional requirements of freshwater zooplankton species. Von Elert et al. (2003) have already suggested that sterols could play a key role in determining carbon transfer efficiency from primary producers to herbivorous zooplankton. Here we suggest that, in addition to low dietary sterol levels, the quality of dietary sterols could strongly affect the assimilation of dietary carbon.

Acknowledgments—We thank A. Wacker for helpful comments on improving the content and style of the manuscript and K.A. Brune for editing the English. This study was supported by the German Research Foundation (DFG, El 179/4-2).

Chapter 4

Trophic upgrading of autotrophic picoplankton food quality by the heterotrophic flagellate *Paraphysomonas* sp.

Alexandre Bec, Dominik Martin-Creuzburg and Eric von Elert

Abstract—This study investigated if trophic repackaging of autotrophic picoplankton by phagotrophic protists is associated with an improvement of food quality for metazoan grazers (trophic upgrading). The nutritional value of different phytoplanktonic species (i.e. *Microcystis aeruginosa* PCC7806, *Synechococcus* sp. BO8809, *Synechococcus elongatus*, *Choricystis minor*) and of the heterotrophic flagellate *Paraphysomonas* sp. grown on these different autotrophic prey items was evaluated in standardized growth experiments with *Daphnia magna*. Results showed that in the simplified autotrophic picoplankton—*Daphnia* food chain the presence of *Paraphysomonas* sp. as an intermediary trophic step enhanced the development and reproduction of *D. magna*. By supplementation of *Synechococcus* sp. with a total lipid extract gained from *Paraphysomonas* sp. (grown on *Synechococcus* sp.) we could show that trophic upgrading of autotrophic picoplankton is caused by the addition of lipids by the heterotrophic flagellate. This trophic upgrading is most probably due to the de novo synthesis of sterols and polyunsaturated fatty acids by *Paraphysomonas* sp.. *Paraphysomonas* sp. also improved the food quality of the toxic strain *M. aeruginosa* PCC7806, which suggests that this heterotrophic flagellate is not only capable of trophic upgrading a poor quality food source by producing essential lipids, but also by detoxifying cyanobacterial food.

Keywords—cyanobacteria, *Synechococcus*, *Microcystis*, *Choricystis*, fatty acids, sterols, toxicity

Introduction

The consumption of picoplanktonic organisms (heterotrophic bacteria, autotrophic picoplankton) by protozoans has been recognized as a major pathway of carbon flow to higher trophic levels (Pomeroy 1974; Azam et al. 1983; Sherr and Sherr 1984). Autotrophic picoplankton (APP) accounts for the bulk of primary production in large parts of open ocean and in many oligotrophic lakes (Li et al. 1983; Stockner and Antia 1986; Weisse 1993; Fogg 1995; Callieri and Stockner 2002). However, such picoplankton is largely unavailable to direct consumption by most crustacean grazers (except cladocerans), whose grazing apparatus is too coarse to retain picoplanktonic particles. In contrast, heterotrophic protists are efficiently grazed by crustacean zooplankton. By repackaging their picoplanktonic prey into particles accessible for crustacean grazers, phagotrophic protists represent a crucial link channelling picoplanktonic production to higher trophic levels (Sherr et al. 1986; Sherr and Sherr 1988; Gifford 1991).

The assimilation of picoplankton by protozoa leads to substantial losses in organic carbon via respiration, which has led to a debate about the quantitative significance of the transfer of picoplanktonic production to higher trophic levels via protists (Ducklow 1986; Sherr et al. 1987). Especially in systems dominated by metazoan grazers that are able to feed directly on picoplankton (i.e. *Daphnia*), trophic repackaging might be regarded as a sink for carbon (Stockner and Shortreed 1989). However, zooplankton production is determined not only by the quantity of the available carbon, but also by its quality. A considerable amount of research has dealt with phytoplankton food quality (Ahlgren et al. 1990; Gulati and DeMott 1997), but surprisingly few studies have focused on the nutritional value of protozoans for metazoan grazers.

In natural systems, especially in freshwater habitats, the availability of phosphorus (Elser et al. 2001; Makino et al. 2002) can determine zooplankton growth. Heterotrophic nanoflagellates (HNFs) and ciliates are rich in phosphorus (Caron and Goldman 1990, Sanders et al. 1996) and, therefore, might be a high quality food source for zooplankton. However, elemental content alone is insufficient to predict food quality. It has been demonstrated that essential lipids, such as (n-3) series polyunsaturated fatty acids (PUFAs) and sterols, can limit zooplankton growth (Müller-Navarra et al. 1995; Müller-Navarra et al. 2000; Wacker and Von Elert 2001; Von Elert 2002; Von Elert et al. 2003; Martin-Creuzburg and Von Elert 2004).

Unfortunately, little is known about the lipid composition of heterotrophic protists. The scarce data available suggest that the lipid composition of protozoans depends on species-specific biosynthetic capacities, but also on the biochemical composition of their food source (Ederington et al. 1995; Desvilettes et al. 1997; Vera et al. 2001; Broglio et al. 2003). This might explain the extreme variability in the observed food quality of protozoans for metazoan grazers (Sanders and Wickham 1993). Several studies have reported that protozoans as a sole food source are of low quality for zooplankton (DeBiase et al. 1990; Sanders et al. 1996; Mohr and Adrian 2002; Bec et al. 2003a), while others argued that HNFs feeding on microalgae are a high quality food for zooplankton (Klein Breteler et al. 1999; Tang et al. 2001; Bec et al. 2003b). These latter studies have also shown that HNFs, as an intermediary trophic step, enhance the nutritional value of their microalgal diet for zooplankton. This trophic upgrading of microalgal food (Klein Breteler et al. 1999; Bec et al. 2003b) has been suggested to be due to new lipid compounds synthesized by the heterotrophic flagellates, but this has not yet been tested.

In this study, we investigated if trophic repackaging of autotrophic picoplankton is associated with an improvement of food quality for metazoans. We used three strains of cyanobacteria (*Microcystis aeruginosa* PCC7806, *Synechococcus* sp. BO8809, *Synechococcus elongatus*) and one picoeukaryotic alga (*Choricystis minor*) as food for the heterotrophic flagellate *Paraphysomonas* sp., and examined if the flagellate as an intermediary trophic step enhanced the development and reproduction of the metazoan grazer *Daphnia magna*. Lipid analyses (fatty acids, sterols) were combined with the supplementation of the flagellate's food source *Synechococcus* sp. with total lipids of *Paraphysomonas* sp. in order to test if trophic upgrading by the heterotrophic flagellate was due to the addition of lipids.

Materials and methods

Cultures and food suspensions of autotrophic organisms—*Synechococcus* sp. strain BO8809 (isolated from Lake Constance by A. Ernst in 1988), *Synechococcus elongatus* (SAG 89.79) and *Microcystis aeruginosa* PCC7806 were grown in Cyano medium (Jüttner et al. 1983) at 20°C with illumination at 60 μmol m^{-2} s^{-1}. *Cryptomonas* sp. (SAG 26.80) and *Choricystis minor* (strain KR1988/8, isolated from Lake Tollensesee by L. Krienitz in 1988) were grown in modified WC medium with vitamins (Guillard 1975) at 20°C with illumination at 120 μmol m^{-2} s^{-1}. The cyanobacteria as well as the algae were cultured semi-continuously at a dilution rate of 0.25 d^{-1} using aerated 5 L vessels. Stock solutions of the different autotrophic organisms for the *Daphnia* growth experiments were prepared by centrifugation and resuspension of the cultured cells in WC without adding vitamins. Carbon concentrations of the autotrophic food suspensions were estimated from photometric light extinction (800 nm) and from carbon-extinction equations previously determined.

Cultures and food suspensions of Paraphysomonas sp.—The chrysomonad *Paraphysomonas* sp. was obtained from the culture collection of the Limnological Institute (University of Konstanz, Germany). By cultivating *Paraphysomonas* sp. on different food organisms (*M. aeruginosa, Synechococcus* sp., *S. elongatus , C. minor*), four flagellate suspensions were obtained, which were used to feed *Daphnia*. These *Paraphysomonas* sp. cultures were grown semi-continuously at 20°C in 500 ml flasks by replacing 10 % of the culture volume with WC medium with vitamins and food every 12 h. Every other day, the *Paraphysomonas* sp. cultures were transferred into new clean flasks. Two days before the *Paraphysomonas* sp. cultures were harvested for the *Daphnia* growth experiments, the supply of medium and food was stopped to allow *Paraphysomonas* sp. to reduce the concentration of autotrophic food particles in the flagellate cultures. However, at the end of this 2-d period, APP cells were still observed as free cells in the culture medium and in the *Paraphysomonas* food vacuoles. For the growth experiments HNF cells were separated from their food source by repeated centrifugation (4 min at 2000×g) and resuspension in fresh medium, taking advantage of their negative geotaxis. Subsequently, the flagellate suspensions were slowly filtered through a 5-μm membrane filter without vacuum, and cells retained on the filter were immediately resuspended in WC without vitamins to obtain protist stock suspensions for the *Daphnia* growth experiments. *Paraphysomonas* sp. cell concentrations were determined with an electronical particle counter and *Paraphysomonas* sp. carbon concentrations offered to daphnids were estimated using the carbon conversion factor proposed by Menden-Deuer and Lessard (2000). Contamination of the flagellate suspensions with heterotrophic bacteria and autotrophic cells was determined according to Porter and Feig (1980). The carbon contributions of heterotrophic bacteria and cyanobacteria were calculated using the conversion factors proposed by Bratbak (1985) and Verity et al. (1992). The *Paraphysomonas* sp. concentrations ranged from 0.7-1.5 mg C L^{-1} and comprised 92-98 % of the total carbon concentration of the food suspensions. The contamination with autotrophic particles never exceeded 4 % of total carbon of either of the *Synechococcus* strains and 6.8 % of *M. aeruginosa* and *C. minor*; the contamination with heterotrophic bacteria was negligible in all treatments (<2 % of the total carbon).

Daphnia growth experiments—Growth experiments were conducted with third clutch juveniles (born within 10 h) of a clone of *Daphnia magna* (Lampert 1991). The experiments were carried out at 20°C in glass beakers filled with 200 ml of filtered lake water (0.45-μm pore-sized membrane filter). Each treatment consisted of three replicates with seven animals each. Somatic growth rates (g) were determined as the increase in dry weight (W) during the experiments using the equation:

$$g = \ln W_t - \ln W_0/t$$

Subsamples of the experimental animals consisting of seven individuals were taken at the beginning (W_0) and at the end (W_t) of an experiment. After 24 h in a drying chamber, subsamples were weighed on an electronic balance (Mettler UMT 2; ± 0.1 μg). Growth rates were calculated as means for each treatment (n = 3).

In a first set of growth experiments we investigated the nutritional value of the autotrophic species (*M. aeruginosa*, *Synechococcus* sp., *S. elongatus*, *C. minor*) and of *Paraphysomonas* sp. grown on each of these autotrophic preys for *D. magna*. A starvation treatment and a high quality food treatment (*Cryptomonas* sp.) served as references. The food suspensions, containing 1.5 mg C L^{-1} of the autotrophic species and 0.7-1.5 mg C L^{-1} of *Paraphysomonas* sp., were renewed daily during the 6-day experiments.

A second set of growth experiments was designed to test if lipids synthesized by *Paraphysomonas* sp. were involved in the trophic upgrading of picocyanobacteria. Therefore, *D. magna* was reared on *Synechococcus* sp., *Synechococcus* sp. supplemented with *Paraphysomonas* sp. lipids and *Synechococcus* sp. supplemented with its own lipids. In a control treatment, *Synechococcus* sp. was supplemented with bovine serum albumin (BSA) without additional lipids to exclude possible effects of BSA on growth of *Daphnia*.

Extraction and supplementation of lipids—Lipids from 20 mg POC of *Paraphysomonas* sp. (grown on *Synechococcus* sp.) or of *Synechococcus* sp. were loaded onto precombusted GF/F filters (Whatman, 47-mm diameter) and extracted three times using a mixture of dichloromethane/methanol (2:1, v/v). The pooled cell-free extracts were evaporated with nitrogen to dryness, weighed to the nearest 0.1 μg and resuspended in ethanol (2.5 mg ml⁻¹). To enrich *Synechococcus* sp. with lipids of *Paraphysomonas* sp. or with its own lipids, 15 mg bovine serum albumin (BSA) was dissolved in 1 ml of ultra-pure water, and 300 μl of the lipid stock solutions were added. After addition of 1 mg particulate organic carbon (POC) of *Synechococcus* sp., the volume was brought to 10 ml with Cyano medium. The resulting suspension was incubated on a rotary shaker (100 rpm min⁻¹) for 4 h. Surplus lipids and BSA were removed by washing the cells three times in 10 ml fresh medium according to Von Elert (2002). The resulting *Synechococcus* sp. suspensions were used as food in the *Daphnia* growth experiments.

Analysis of fatty acids and sterols—For the analysis of fatty acids and sterols, the food suspensions (sampled on day 6) were filtered on precombusted GF/F filters (Whatman, 25 mm diameter). POC was measured using a NCS-2500 analyzer (Carlo Erba Instruments). The total lipids were extracted with dichloromethane/methanol as described above and evaporated to dryness with nitrogen. The lipid extracts were transesterified with 3 mol L^{-1} methanolic HCl (60°C, 15 min) for the analysis of fatty acids or saponified with 0.2 mol L^{-1} methanolic KOH (70°C, 1 h) for the analysis of sterols. Subsequently, fatty acid methyl esters (FAMEs) were extracted three times with 2 ml of *iso*-hexane, the neutral lipids (sterols) were partitioned into *iso*-hexan:diethyl ether (9:1, v/v). The lipid fractions were evaporated to dryness under nitrogen and resuspended in a volume of 10-20 μl of *iso*-hexane. Lipids were analyzed by gas chromatography on an HP 6890 GC equipped with a flame ionization detector and a DB-225 (J&W Scientific) capillary column to analyze fatty acids or with an HP-5 (Agilent) capillary column to analyze sterols, as described by Von Elert (2002) for fatty acids and Martin-Creuzburg and Von Elert (2004) for sterols. Lipids were quantified by comparison to internal standards (C17:0 and C23:0 methyl esters; 5α-cholestan) and identified by their retention times and their mass spectra, which were recorded with a gas chromatograph-mass spectrometer (Finnigan MAT GCQ) equipped with a fused-silica capillary column (DB-225MS, J&W for FAMEs; DB-5MS, Agilent for sterols). Sterols were analyzed as free sterols and as their steryl acetate derivatives. Spectra were recorded between 50 and 600 amu in the EI ionization mode. Mass spectra were identified by comparison with mass spectra of reference substances (purchased from Sigma, Supelco or Steraloids) or spectra found in the literature. A clear positive proof of the identification of the C24 epimers poriferasterol/stigmasterol as well as of the cis/trans isomers isofucosterol/fucosterol could not be achieved. The absolute amount of each lipid was related to the particulate organic carbon (POC), which was determined from an additional aliquot of the food suspensions used for lipid analysis.

Data analysis—Growth rates and clutch sizes were analyzed by one-way analyses of variance (ANOVA) and post hoc comparisons (Tukey's HSD, $\alpha = 0.05$). Raw data met the assumption of homogeneity of variance (Levene's test).

Results

Growth of Daphnia magna—Juvenile somatic growth rates (g) of *D. magna* were significantly affected by the food supplied in the growth experiments (ANOVA, $F_{8,18} = 649$; $P < 0.001$; Fig. 1). In the bitrophic systems growth rates of *D. magna* ranged from -0.06 d^{-1} (starving) to 0.51 d^{-1} (*Cryptomonas* sp.). Somatic growth of *D. magna* on both *Synechococcus* strains was poor. However, when *Paraphysomonas* sp. was introduced as an intermediary trophic level between the picocyanobacterial strains and the metazoan grazer, growth of *D. magna* was increased 7-8-fold, which indicated a significant improvement in cyanobacterial food quality by *Paraphysomonas* sp. (Tukey's HSD, $P < 0.05$).

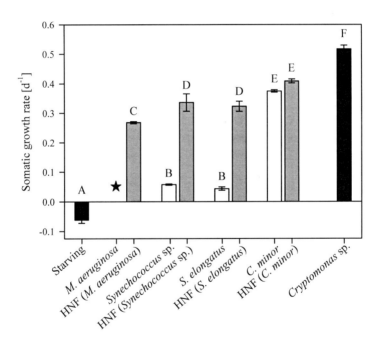

Fig. 1 Juvenile somatic growth rates of *Daphnia magna* grown on *Microcystis aeruginosa*, *Synechococcus* sp., *Synechococcus elongatus*, *Choricystis minor*, and on the heterotrophic flagellate (HNF) *Paraphysomonas* sp. fed with either of these autotrophic species. Growth without food supply (starving) and growth on *Cryptomonas* sp. are shown for comparison. Daphnids grown on the toxic *M. aeruginosa* did not survive the 6-day experiment (indicated by an asterisk). Data are means of n = 3; error bars indicate SD. Bars labelled with the same letters are not significantly different (Tukey's HSD, P < 0.05 following ANOVA).

Daphnids fed on the toxic *M. aeruginosa* strain did not survive the 6-day experiments. In contrast, daphnids fed on *Paraphysomonas* sp. grown on *M. aeruginosa* all survived and grew relatively well. Growth of *D. magna* on *C. minor* was substantially higher than on picocyanobacteria (Tukey's HSD, $P < 0.05$; Fig.1), but did not significantly increase when *Paraphysomonas* sp. was interposed as intermediary grazer (Tukey's HSD, $P = 0.10$).

Clutch sizes of Daphnia magna—*D. magna* did not produce eggs when grown on *Synechococcus* sp., *S. elongatus* or *Paraphysomonas* sp. fed with *M. aeruginosa* or in the starving treatment (Fig. 2).

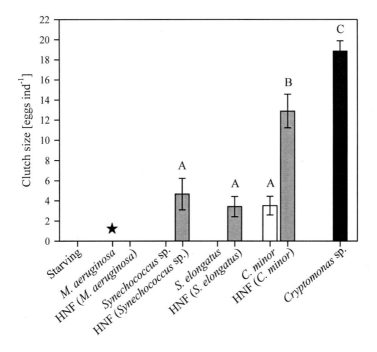

Fig. 2 Clutch sizes of *Daphnia magna* grown on *Microcystis aeruginosa*, *Synechococcus* sp., *Synechococcus elongatus*, *Choricystis minor*, and on the heterotrophic flagellate (HNF) *Paraphysomonas* sp. fed with either of these autotrophic species. The clutch size of *D. magna* grown on *Cryptomonas* sp. is shown for comparison. Daphnids grown on the toxic *M. aeruginosa* did not survive the 6-day experiment (indicated by an asterisk). Data are means of $n = 3$; error bars indicate SD. Bars labelled with the same letters are not significantly different (Tukey's HSD, $P < 0.05$ following ANOVA).

When *Paraphysomonas* sp. was interposed between either of the two *Synechococcus* strains and *D. magna*, the animals exhibited intermediary clutch sizes. Daphnids fed on *C. minor* produced comparatively small clutches (Fig. 2); however, the clutch sizes increased significantly when *Paraphysomonas* sp. was introduced as an intermediary grazer (Tukey's HSD, $P < 0.05$ following ANOVA, $F_{4,10} = 87.9$; $P < 0.001$). The largest clutch sizes were produced when *Cryptomonas* sp. was offered as food.

Effects of lipid supplementation on growth of Daphnia—The supplementation of *Synechococcus* sp. with BSA did not affect *Daphnia* growth (Tukey's HSD following ANOVA, $F_{4,10} = 361$; $P < 0.001$; Fig. 3). Similarly, supplementation of *Synechococcus* sp. with its own lipids did not improve growth of the grazer, which indicates that the food quality of *Synechococcus* sp. was not due to a shortage of lipids in general. When *Synechococcus* sp. was supplemented with lipids from *Paraphysomonas* sp., growth of *D. magna* increased significantly (Tukey's HSD, $P < 0.05$; Fig. 3), which indicates that the low food quality of *Synechococcus* sp. was due to the absence of a lipid which is present in the flagellate.

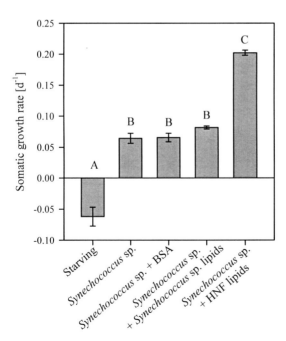

Fig. 3 Juvenile somatic growth rates of *Daphnia magna* grown on *Synechococcus* sp. and on *Synechococcus* sp. supplemented with BSA, a total lipid extract gained from *Synechococcus* sp. and a total lipid extract gained from the heterotrophic flagellate (HNF) *Paraphysomonas* sp. which itself previously fed on *Synechococcus* sp.. Data are means of n = 3; error bars indicate SD. Bars labelled with the same letters are not significantly different (Tukey's HSD, P < 0.05 following ANOVA).

Sterol compositions—No sterol were detected in any of the cyanobacterial strains. Stigmasterol (24α-ethylcholesta-5,22-dien-3β-ol) and epibrassicasterol (24α-methylcholesta-5,22-dien-3β-ol) were the principal sterols in *Cryptomonas* sp. (Tab. 1). Epibrassicasterol, the 24α-epimer of brassicasterol, has previously been reported to occur in cryptophycean algae such as *Cryptomonas* and *Rhodomonas* (Goad et al. 1983; Gladu et al. 1990). Although we did not determine the side chain stereochemistry at C-24, the presence of epibrassicasterol rather than brassicasterols in the *Cryptomonas* sp. strain we used here was also assumed. Three major sterols were detected in the green alga *C. minor*: fungisterol (24β-methylcholesta-7-en-3β-ol), ergosterol (24β-methylcholesta-5,7,22-trien-3β-ol) and chondrillasterol (24β-ethylcholesta-7,22-dien-3β-ol), all of them common in chlorophycean species (e.g. Rezanka et al. 1986; Cranwell et al. 1990; Martin-Creuzburg et al. 2005).

The chrysomonad *Paraphysomonas* sp. contained 24-ethylcholesta-5,22-dien-3β-ol as the major sterol. The two C-24 epimers of 24-ethylcholesta-5,22-dien-3β-ol, poriferasterol (24β) and stigmasterol (24α) could not be separated definitely on the GC-FID and GC-MS systems used, so that the stereochemistry of the ethylgroup at C-24 was not defined. The presence of poriferasterol in the related chrysomonad *Ochromonas* (Gershengorn et al. 1968) suggests that the 24β-configuration of 24-ethylcholesta-5,22-dien-3β-ol is also present in *Paraphysomonas* sp.. 24-Ethylidenecholesta-5,24(28)-dien-3β-ol was the second sterol found in *Paraphysomonas* sp.. To assess the configuration of the C24 ethylidene group, which can either be cis (24Z, isofucosterol) or trans (24E, fucosterol), sterols were analyzed as their steryl acetates. The mass spectra of the acetates showed a dominant m/z 296 ion and a much less abundant m/z 394 ion, which indicates the presence of isofucosterol rather than fucosterol (Knights and Brooks 1969).

The sterol composition of *Paraphysomonas* sp. did not change with its food source; i.e. *Paraphysomonas* sp. grown on a sterol free cyanobacterial diet exhibited the same sterol pattern as *Paraphysomonas* sp. grown on the sterol containing green alga *C. minor* (Tab. 1). Only trace amounts of the sterols found in *C. minor* were detected in *Paraphysomonas* sp. grown on this eukaryotic alga. These traces could be attributed to the slight contamination of the flagellate suspensions with *C. minor* cells and/or to undigested material in the flagellate food vacuoles. The total sterol content of *Paraphysomonas* sp. was generally higher than the total sterol content of the *C. minor* and *Cryptomonas* sp. (Tab. 1).

Table 1. Sterol content of the heterotrophic nanoflagellate (HNF) *Paraphysomonas* sp. grown on different food sources (*Microcystis aeruginosa*, *Synechococcus* sp., *Synechococcus elongatus*, or *Choricystis minor*) and of *C. minor* and *Cryptomonas* sp.. No sterols were detected in the two *Synechococcus* strains and in *M. aeruginosa*. The values are means of n = 3; ± SD.

Sterol	HNF (*M. aeruginosa*) [µg mg C⁻¹]	[%]	HNF (*Synechococcus* sp.) [µg mg C⁻¹]	[%]	HNF (*S. elongatus*) [µg mg C⁻¹]	[%]	HNF (*C. minor*) [µg mg C⁻¹]	[%]	*C. minor* [µg mg C⁻¹]	[%]	*Cryptomonas* sp. [µg mg C⁻¹]	[%]
Poriferasterol	10.9 ± 0.1	85.0	8.0 ± 0.2	75.7	8.4 ± 1.0	66.7	6.0 ± 0.8	70.2	–		–	
Isofucosterol	1.9 ± 0.0	15.0	2.6 ± 0.7	24.3	4.2 ± 0.1	33.3	2.6 ± 0.1	29.8	–		–	
Fungisterol	–		–		–		trace		2.3 ± 0.1	45.9	–	
Ergosterol	–		–		–		trace		1.7 ± 0.5	33.9	–	
Chondrillast.	–		–		–		–		1.0 ± 0.1	20.2	–	
Stigmasterol	–		–		–		–		–		4.0 ± 0.2	57.7
Epibrassicast.	–		–		–		–		–		2.9 ± 0.1	42.3
Total	12.8 ± 0.1	100	10.6 ± 0.6	100	12.6 ± 0.9	100	8.6 ± 0.8	100	5.0 ± 0.7	100	6.9 ± 0.3	100

Fatty acid compositions—The fatty acid composition of both *Synechococcus* strains was characterized by high amounts of 16:0 and 16:1(n-7), and by the lack of PUFAs (Tab. 2). In contrast, *Microcystis* was characterized by significant amounts of (n-6) series PUFAs mainly represented by 18:2(n-6) and 18:3(n-6), which is in accordance with Gugger et al. (2002). Like other Chlorophyceae (Ahlgren et al. 1992; Von Elert and Wolffrom 2001), *C. minor* exhibited significant amounts of 16:3(n-3), 18:2(n-6) and 18:3(n-3). *Cryptomonas* sp. was characterized by a clear dominance of (n-3) series PUFAs, mainly 18:3(n-3), 18:4(n-3) and 20:(5n-3).

The fatty acid composition of *Paraphysomonas* sp. grown on the different autotrophic species exhibited significant amounts of 20:4(n-6), 20:5(n-3) and 22:6(n-3) (Tab. 3). Since both *Synechococcus* strains lack (n-3) series PUFAs, this demonstrated that *Paraphysomonas* sp. is able to synthesize these long-chain PUFAs de novo. Autotrophic chrysophytes are known to produce long-chain PUFAs in high quantities (Ahlgren et al. 1992), which suggests that the capacity of *Paraphysomonas* sp. to synthesize PUFAs de novo could be attributed to its phylogenetic origin. However, this should be relativised as Vera et al. (2001) reported that *Paraphysomonas vestita* cultivated on heterotrophic bacteria exhibited only very low amounts of (n-3) series PUFA. Nevertheless, Zhukova and Kharlamenko (1999) showed that the flagellate *Bodo* sp. exhibited great variations in its (n-3) series PUFA content when fed on rice grown bacteria or on starch-grown bacteria.

The fatty acid composition of *Paraphysomonas* sp. was furthermore influenced by its diet. Like *C. minor*, *Paraphysomonas* sp. grown on this picoeukaryotic algae is characterized by significant amounts of 16:3(n-3), 18:2(n-6) and 18:3(n-3). This result is in good agreement with a previous study which reported that the heterotrophic flagellate *Aulacomonas submarina* accumulated C_{16} and C_{18} PUFAs when fed on the chlorophycea *Chlorogonium elongatum* (Bec et al. 2003b). *C. minor* showed a higher content of 18:2(n-6) than of 18:3(n-3) (Tab. 2). However, this ratio is inversed in *Paraphysomonas* sp. when feeding on *C. minor* (Tab. 3), which indicates a preferential assimilation of 18:3(n-3) by the flagellate.

Table 2. Fatty acid content and composition of *Microcystis aeruginosa*, *Synechococcus* sp., *Synechococcus elongatus*, *Choricystis minor* and *Cryptomonas* sp. Values are given as means of n = 3; ± SD. n.d. = not detected.

Fatty acid	*M. aeruginosa* [µg mg C⁻¹]	[%]	*Synechococcus* sp. [µg mg C⁻¹]	[%]	*S. elongatus* [µg mg C⁻¹]	[%]	*C. minor* [µg mg C⁻¹]	[%]	*Cryptomonas* sp. [µg mg C⁻¹]	[%]
16:0	47.8 ± 0.2	36.7	12.3 ± 0.4	16.4	24.0 ± 0.8	26.2	40.7 ± 0.4	15.2	20.7 ± 0.2	9.0
16:1n-7	6.1 ± 0.1	4.7	34.4 ± 1.6	46.0	46.9 ± 1.9	51.3	4.0 ± 0.1	1.5	4.9 ± 0.1	2.2
16:3n-3	n.d.	—	n.d.	—	n.d.	—	23.5 ± 0.4	8.8	n.d.	—
17:1n-7	0.8 ± 0.0	0.6	0.6 ± 0.0	0.8	0.5 ± 0.0	0.5	n.d.	—	1.1 ± 0.0	0.5
18:0	12.9 ± 0.1	9.9	1.7 ± 0.0	2.3	1.5 ± 0.0	1.6	2.7 ± 0.0	1.0	3.5 ± 0.0	1.5
18:1n-9	16.8 ± 0.2	12.9	0.8 ± 0.0	1.0	0.7 ± 0.1	0.7	21.0 ± 0.3	7.8	4.7 ± 0.1	2.1
18:1n-7	5.8 ± 0.1	4.5	1.6 ± 0.0	2.1	1.8 ± 0.0	1.9	1.7 ± 0.7	0.6	7.0 ± 0.0	3.1
18:2n-6	20.2 ± 0.2	15.5	n.d.	—	n.d.	—	99.6 ± 1.2	37.2	16.5 ± 0.1	7.2
18:3n-6	14.4 ± 0.1	11.1	n.d.	—	n.d.	—	n.d.	—	1.1 ± 0.0	0.5
18:3n-3	0.5 ± 0.0	0.4	n.d.	—	n.d.	—	70.5 ± 0.8	26.4	74.5 ± 0.3	32.6
18:4n-3	n.d.	—	n.d.	—	n.d.	—	n.d.	—	48.0 ± 0.1	21.0
20:1n-9	n.d.	—	n.d.	—	n.d.	—	n.d.	—	1.6 ± 0.1	0.7
20:2n-6	n.d.	—	n.d.	—	n.d.	—	n.d.	—	n.d.	—
20:4n-6	n.d.	—	n.d.	—	n.d.	—	n.d.	—	0.9 ± 0.0	0.4
20:3n-3	n.d.	—	n.d.	—	n.d.	—	n.d.	—	0.8 ± 0.0	0.4
20:5n-3	n.d.	—	n.d.	—	n.d.	—	n.d.	—	34.3 ± 0.1	15.0
22:6n-3	n.d.	—	n.d.	—	n.d.	—	n.d.	—	3.1 ± 0.0	1.4
Total	130.2	100	74.7	100	91.5	100	267.5	100	228.4	100
Total (n-3)	0.5	0.4	n.d.	—	n.d.	—	94.7	28.5	160.8	68.4

Table 3. Fatty acid content and composition of the heterotrophic flagellate (HNF) *Paraphysomonas* sp. grown on *Microcystis aeruginosa*, *Synechococcus* sp., *Synechococcus elongatus* or *Choricystis minor* respectively. Values are given as means of n = 3; ± SD. n.d. = not detected.

Fatty acid	HNF (*M. aeruginosa*) [µg mg C^{-1}]	[%]	HNF (*Synechococcus* sp.) [µg mg C^{-1}]	[%]	HNF (*S. elongatus*) [µg mg C^{-1}]	[%]	HNF (*C. minor*) [µg mg C^{-1}]	[%]
16:0	11.1 ± 0.6	15.8	12.0 ± 0.7	14.2	15.8 ± 0.0	16.9	11.6 ± 0.1	15.2
16:1n-7	10.7 ± 0.8	15.2	21.3 ± 1.3	25.3	20.1 ± 0.1	21.6	3.7 ± 0.1	4.8
16:3n-3	n.d.	—	n.d.	—	n.d.	—	2.9 ± 0.0	3.8
17:1n-7	2.4 ± 0.1	3.4	1.5 ± 0.0	1.8	1.4 ± 0.0	1.5	1.4 ± 0.0	1.8
18:0	3.4 ± 0.1	4.8	2.8 ± 0.1	3.3	3.4 ± 0.1	3.7	4.2 ± 0.1	5.6
18:1n-9	4.1 ± 0.1	5.8	2.5 ± 0.1	2.9	4.5 ± 0.2	4.8	5.3 ± 0.1	7.0
18:1n-7	13.2 ± 0.4	18.8	15.9 ± 0.6	18.8	16.3 ± 0.1	17.5	6.2 ± 0.2	8.1
18:2n-6	2.8 ± 0.1	4.0	3.3 ± 0.0	3.9	3.8 ± 0.1	4.1	11.2 ± 0.0	14.7
18:3n-6			1.8 ± 0.0	2.1	1.6 ± 0.0	1.7	0.3 ± 0.1	0.4
18:3n-3	1.3 ± 0.0	1.8	1.6 ± 0.0	1.9	1.5 ± 0.0	1.6	13.7 ± 0.4	18.0
18:4n-3	2.1 ± 0.0	3.0	n.d.	—	n.d.	—	1.2 ± 0.1	1.6
20:1n-9	n.d.	—	n.d.	—	n.d.	—	n.d.	—
20:2n-6	n.d.	—	n.d.	—	n.d.	—	1.2 ± 0.0	1.5
20:4n-6	7.2 ± 0.1	10.2	4.4 ± 0.0	5.2	6.4 ± 0.3	6.9	1.6 ± 0.1	2.1
20:3n-3	n.d.	—	n.d.	—	0.8 ± 0.4	0.8	1.4 ± 0.0	1.8
20:5n-3	3.1 ± 0.1	4.4	3.3 ± 0.0	3.9	5.6 ± 0.2	6.0	2.6 ± 0.0	3.4
22:6n-3	4.2 ± 0.1	5.9	6.0 ± 0.1	7.2	6.7 ± 0.0	7.2	2.6 ± 0.1	3.4
Total	70.4	100	84.3	100	93.3	100	76.2	100
Total (n-3)	10.7	15.0	10.9	12.7	14.5	15.2	24.4	30.1

Discussion

Autotrophic picoplankton (APP) is the major primary producer in many oligo- to mesotrophic marine and freshwater ecosystems. In marine systems, where copepods are the major grazers, this production becomes available only after assimilation by protozoa as an intermediary trophic level. This trophic situation differs from freshwater systems, where zooplankton is often dominated by cladocerans of the genus *Daphnia*. Daphnids can either feed directly on APP or gain indirect access to APP production when they consume protists feeding on APP. Here, we investigated these two contrasting situations for daphnids in simplified bi- and tritrophic food chains. In the different *Paraphysomonas* sp. treatments, dietary carbon provided to daphnids consisted to more than 92 % of the heterotrophic flagellate, which allowed us to clearly assign effects of the *Paraphysomonas* sp. food suspension to growth of *D. magna*. Although in the different *Paraphysomonas* sp. treatments daphnids were offered a lower quantity of food than those feeding on the different autotrophic species, food quantity was well above the incipient limiting level (Lampert 1977), which allowed us to interpret differences in growth and reproduction of *D. magna* as an effect of food quality and not of food quantity.

Growth rates and clutch sizes of *D. magna* on prokaryotic APP were low, which is in accordance with the well known low food quality of cyanobacteria for daphnids (Von Elert and Wollfrom 2001; Von Elert et al. 2003). The introduction of *Paraphysomonas* sp. as an intermediary trophic level led to a 7-8-fold increase of somatic growth of *D. magna* and to significantly increased clutch sizes. Hence, the heterotrophic nanoflagellate (HNF) *Paraphysomonas* sp. clearly upgraded the food quality of autotrophic picocyanobacteria.

When *Paraphysomonas* sp. was interposed in the *C. minor*—*Daphnia* food chain, effects on growth and reproduction of *D. magna* differed: the presence of *Paraphysomonas* sp. did not affect somatic growth but significantly increased the clutch size of the cladoceran. The green alga *C. minor* contained comparatively high amounts of sterols, but was deficient in 20:5(n-3), a potentially essential PUFA for *Daphnia*. This emphasized the importance of sterols for somatic growth of *Daphnia* and suggested that certain PUFAs are rather needed for reproduction. Although it is generally assumed that effects of food quantity on somatic growth and on reproduction are highly correlated in juvenile *Daphnia* (Lampert and Trubetskova 1996), it has previously been suggested that limitation by food quality might affect somatic growth and reproduction differently. This has been shown for mineral (Urabe and Sterner 2001) and biochemical (Becker and Boersma 2003; Martin-Creuzburg and Von Elert 2004) aspects of food limitation.

It should also be noted that growth and reproduction of *D. magna* differed significantly between the *Paraphysomonas* sp. treatments, which indicates that the nutritional value of *Paraphysomonas* sp. varies with its picoplanktonic prey. However, the nutritional value of *Paraphysomonas* sp. for *D. magna* was comparatively high in all treatments. Therefore, effects of the *Paraphysomonas* sp. autotrophic prey on the flagellates food quality for crustacean grazers are of minor importance.

It has been hypothesized that trophic upgrading of poor quality microalgae is due to the the addition of essential lipids by the intermediary heterotrophic flagellates (Klein Breteler et al. 1999; Bec et al. 2003b). This hypothesis is corroborated by our finding that supplementation of *Synechococcus* sp. with a total lipid extract gained from *Paraphysomonas* sp. (grown on this *Synechococcus* strain) significantly enhanced *Daphnia* growth, which strongly suggests that trophic upgrading by *Paraphysomonas* sp. is due to the addition of essential lipids.

In contrast to *Paraphysomonas* sp., the three cyanobacterial strains we used here did not contain sterols. Sterols are involved in a wide range of physiological processes in arthropods. However, arthropods are incapable of synthesizing sterols de novo, and, therefore, must obtain these essential nutrients from their diet (Goad 1981). It has recently been shown that the absence of sterols in cyanobacteria is the major food quality constraint for daphnids (Von Elert et al. 2003; Martin-Creuzburg and Von Elert 2004). Therefore, the observed trophic upgrading of picocyanobacterial food is most likely due to the addition of sterols to the dietary carbon. This is supported by the high quantities of sterols in *Paraphysomonas* sp. ranging from 8.6 µg mg C^{-1} (*Paraphysomonas* sp. grown on *C. minor*) to 12.8 µg mg C^{-1} (*Paraphysomonas* sp. grown on *M. aeruginosa*) that exceed even those of the eukaryotic algae (*Cryptomonas* sp., *C. minor*). Moreover, *Paraphysomonas* sp. sterols are predominated by Δ^5 sterols, which have been shown to support somatic growth and reproduction of daphnids when supplemented to the sterol-free *S. elongatus* (Martin-Creuzburg and Von Elert 2004).

In general, cyanobacteria do not only lack sterols (Volkman 2003), most of them are also deficient in long-chain PUFAs. Several studies have underlined the importance of (n-3) series PUFAs, in particular of long-chain PUFAs such as 20:5(n-3), for growth and reproduction of cladocerans (Ahlgren et al. 1990; Brett & Müller-Navarra 1997; Weers & Gulati 1997; Park et al. 2002; Bec et al. 2003a). However, Von Elert et al. (2003) have shown that daphnids are primarily sterol limited when grown on cyanobacteria and that PUFAs become limiting only when the shortage of sterols has been overcome by sterol supplementation. *Paraphysomonas* sp. differed from its picocyanobacterial prey in the presence of both sterols and PUFAs. Provided that the high sterol content of *Paraphysomonas* sp. fully compensated for the cyanobacterial sterol deficiency, PUFAs present in *Paraphysomonas* sp. might have further contributed to the trophic upgrading of the picocyanobacteria.

Trophic upgrading by *Paraphysomonas* sp. was also observed for the green alga *C. minor*, though this effect was detectable only for reproduction and not for somatic growth of *D. magna*. The lipid composition of *C. minor* differed from that of *Paraphysomonas* sp. by the predominance of Δ^7-sterols, which are less efficiently used by *Daphnia* (Martin-Creuzburg and Von Elert 2004), and by the absence of 20:5n-3. This suggests that the increase in clutch sizes of *D. magna*, when *Paraphysomonas* sp. was interposed as an intermediary grazer, is either due to the addition of Δ^5-sterols or to the addition of long-chain (n-3) series PUFAs. Martin-Creuzburg et al. (2005) suggested that, in contrast to somatic growth, reproduction is less sensitive to sterol limitation. Thus, the increased cluch size is most likely caused by the addition of PUFAs, such as 20:5n-3, by the interposed flagellate. Trophic upgrading of the green alga *C. minor* might therefore be assigned mainly to the synthesis of long-chain (n-3) series PUFAs, while the amelioration of picocyanobacterial carbon by *Paraphysomonas* sp. is caused by the synthesis of both sterols and (n-3) series PUFAs.

In case of *Paraphysomonas* sp. grown on *M. aeruginosa*, other factors than the lipid metabolism should be considered. *M. aeruginosa* PCC7806 has clearly been shown to be toxic for *Daphnia* (Jungmann 1992; Lürling et al. 2003), which could be confirmed in this study when *D. magna* was feeding directly on *M. aeruginosa*. However, when daphnids fed on *Paraphysomonas* sp. that had been raised on *M. aeruginosa*, no mortality was observed and *D. magna* exhibited comparatively high growth rates. In contrast to growth of *Daphnia*, growth of *Paraphysomonas* sp. was not affected by *M. aeruginosa* toxicity. Several phagotrophic protists are known to graze and grow efficiently on toxic phytoplankton species (Jeong et al. 2001; Stoecker 2002) and some have been suggested to reduce the toxicity of their prey for copepods (Jeong et al. 2001, 2003). In accordance with these findings, we show here that *Paraphysomonas* sp. as an intermediary grazer is capable of detoxifying cyanobacterial food for *Daphnia*. In comparison to growth and reproduction of *D. magna* on *Paraphysomonas* sp. fed on non-toxic picocyanobacteria, animals grown on *Paraphysomonas* sp. which had been raised on *M. aeruginosa* showed significantly reduced growth rates and did not produce eggs. This might be attributed to a remaining toxicity due to unremoved *M.*

aeruginosa cells in the daphnids' diet or to undigested material in the food vacuoles of the flagellate.

In conclusion, we provided evidence that heterotrophic flagellates are capable of biochemically upgrading prokaryotic and eukaryotic picoplankton for metazoan grazers by producing sterols and PUFAs. We furthermore demonstrated that the heterotrophic flagellate *Paraphysomonas* sp. is capable of detoxifying cyanobacterial food for *Daphnia*. Although phagotrophic protists as intermediary grazers represent an additional trophic step at which energy is lost by respiration, our data strongly emphasize that heterotrophic flagellates provide a crucial link between poor quality primary producers (e.g. picocyanobacteria) and higher trophic levels.

Chapter 5

Trophic upgrading of picocyanobacterial carbon by ciliates for nutrition of *Daphnia magna*

Dominik Martin-Creuzburg, Alexandre Bec and Eric von Elert

Abstract—Unicellular picocyanobacteria, such as species of the genus *Synechococcus*, are unsuitable for supporting growth and reproduction of the aquatic key-stone species of the genus *Daphnia*. In *Synechococcus*, long-chain polyunsaturated fatty acids (PUFAs) and sterols are absent, which leads to a low carbon transfer efficiency at the picocyanobacteria-*Daphnia* interface. Here we addressed the question whether ciliates can serve as a trophic link between cyanobacterial APP production and *Daphnia* production, thereby upgrading the nutritional value of an APP food source by producing essential lipids such as PUFAs or sterols. In simplified experimental food chains, consisting of one of two different *Synechococcus* strains, the ciliate *Colpidium campylum* or *Cyclidium* sp., and *Daphnia magna*, we provided evidence that predation on ciliates by *Daphnia* allows access to APP production. Since daphnids are primarily sterol-limited when grown on the picocyanobacterium *Synechococcus*, the observed trophic upgrading of APP food quality by intermediary ciliates is most probably due to the addition of sterols or sterol-like compounds that at least partly release *Daphnia* from sterol limitation. The absence of sterols in ciliates suggests that tetrahymanol and/or hopanoids provide functional equivalents of sterols not only in ciliates but also in *Daphnia*, thereby leading to enhanced growth of the cladocerans.

Keywords—*Synechococcus*, *Colpidium*, *Cyclidium*, sterols, tetrahymanol, hopanoids

Introduction

The significance of autotrophic picoplankton (APP) for primary production in aquatic ecosystems has often been recognized. Unicellular picocyanobacteria are the major constituents of APP; they often dominate both the total phytoplankton biomass and the production in oligotrophic to mesotrophic lakes (Weisse 1993, Callieri & Stockner 2002). Numerous planktonic organisms are able to feed on picoplankton-sized particles. Heterotrophic nanoflagellates and small ciliates are considered to be the most important APP grazers (Sanders et al. 1989, Weisse et al. 1990, Šimek et al. 1995, Pernthaler et al. 1996), and contribute themselves significantly to zooplankton nutrition, thus transferring energy via the microbial loop to higher trophic levels (Stoecker & McDowell Capuzzo 1990). However, the importance of protozoa for the carbon transfer efficiency in aquatic food webs is controversial. By repackaging their prey into accessible particles, protozoa are often implicated as a 'trophic link' to metazoan grazers (Gifford 1991). Copepods prey inefficiently on small particles so that an intermediate protozoan level might give access to APP production that otherwise would not be available for higher trophic levels (Jack & Gilbert 1993). In contrast, both APP and small ciliates fall within the prey size spectrum of the aquatic key-stone species of the genus *Daphnia* (Jürgens 1994) and, therefore, an intermediate trophic level can also be considered as a 'sink' of energy since the efficiency with which carbon is transferred to higher trophic levels depends mainly on the number of trophic steps it has to pass through (Pomeroy & Wiebe 1988). However, picocyanobacteria, such as species of the genus *Synechococcus*, are unsuitable for supporting *Daphnia* growth and reproduction (Lampert 1977*a,b*, 1981, Von Elert et al. 2003, Martin-Creuzburg & Von Elert 2004, Martin-Creuzburg et al. 2005); therefore, cladocerans might also benefit from grazing on protozoa, provided that protozoan carbon is more suitable for *Daphnia*.

In *Synechococcus* long-chain polyunsaturated fatty acids (PUFAs) and sterols are absent, which led to the hypothesis that the low carbon transfer efficiency at the cyanobacteria-*Daphnia* interface is caused by the lack of either of these essential lipid classes (DeMott & Müller-Navarra 1997, Von Elert & Wolffrom 2001, Von Elert et al. 2003). Therefore, the role of protozoa as intermediary grazers might not be restricted to channelling energy; they might also upgrade the biochemical composition of food deficient in essential lipids by producing PUFAs or sterols (Klein Breteler et al. 1999, Bec et al. 2003*b*, Tang & Taal 2005).

Ciliates are abundant protists in freshwater ecosytems (Pace & Orcutt 1981, Beaver & Crisman 1982, Pace 1982, Porter et al. 1985, Müller 1989). Reports from laboratory (Tezuka 1974, Porter et al. 1979, Sanders et al. 1996) and field experiments (Carrick et al. 1991, Pace

& Funke 1991, Marchessault & Mazumder 1997, Jürgens et al. 1999, Zöllner et al. 2003) have revealed that cladocerans are important ciliate predators. However, investigations of food quality aspects of ciliates in *Daphnia* nutrition are scarce. Available data suggest that, albeit rich in nitrogen and phosporus, ciliates are less nutritious for daphnids than many algae (DeBiase et al. 1990, Sanders et al. 1996). The fatty acid composition of ciliates seems to be highly diverse and points to species-specific differences in the capacity to synthesize PUFAs potentially important for zooplankton growth (Kaneshiro et al. 1979, Desvilettes et al. 1997, Sul et al. 2000, Klein Breteler et al. 2004). Although, ciliates presumably lack the ability to synthesize sterols de-novo, exogenously supplied sterols can be incorporated into cell membranes and metabolized into various other sterols (Conner et al. 1968, Harvey & McManus 1991, Harvey et al. 1997). In the absence of exogenous sterols, the pentacyclic triterpenoid alcohol tetrahymanol is produced; tetrahymanol is functionally equivalent to sterols as a structural component of cell membranes in ciliates (Conner et al. 1968). Ederington et al. (1995) reported the assimilation of tetrahymanol in copepod tissues when ciliates were offered as food, but possible effects of tetrahymanol on growth and reproduction of zooplankton with regard to sterol limitation have not yet been identified.

In the present study, we investigated the potential of ciliates to serve as a trophic link between APP and zooplankton production. In experimental tritrophic food chains, consisting of one of two different *Synechococcus* strains, the ciliate *Colpidium campylum* or *Cyclidium* sp., and *Daphnia magna*, we tested the ability of ciliates as intermediary grazers to upgrade the food quality of cyanobacterial APP for *Daphnia* by producing essential lipids such as PUFAs or sterols.

Materials and methods

Cultivation of cyanobacteria and algae—The green alga *Scenedesmus obliquus* (SAG 276-3a, Sammlung von Algenkulturen Göttingen, Germany) was used as food for stock cultures of *Daphnia magna*; it was grown in batch cultures in Cyano medium at 20°C (Jüttner et al. 1983) and harvested in the late-exponential phase. The coccoid cyanobacteria *Synechococcus elongatus* (SAG 89.79) and *Synechococcus* sp. strain BO8809 (isolated from Lake Constance by A. Ernst in 1988, for detailed information see Ernst et al. 1991) were each grown in Cyano medium at 20°C with illumination at 60 µmol m^{-2} s^{-1}. *Cryptomonas* sp. (SAG 26.80) was grown in modified WC medium containing vitamins (Guillard 1975) at 20°C with illumination at 120 µmol m^{-2} s^{-1}. The cyanobacteria and *Cryptomonas* sp. were cultured semi-continuously at a dilution rate of 0.25 d^{-1} in aerated 5-l vessels. Stock solutions of these organisms for the growth experiments were prepared by concentrating the cells by centrifugation and resuspension in WC medium lacking vitamins. Carbon concentrations of the food suspensions were estimated from photometric light extinction (800 nm) and from carbon-extinction equations determined previously.

Cultivation of protozoa—*Colpidium campylum* was obtained from the Laboratoire de Biologie des Protistes (Université Blaise Pascal, France) and *Cyclidium* sp. was obtained from the culture collection of the Limnological Institute (University of Konstanz, Germany). The protozoa were cultivated semi-continuously at 20°C in mineral water (Volvic®). Ciliates were fed with one of the two *Synechococcus* strains at approximatively 2 mg C l^{-1}, and, in order to maintain exponential growth, 20–40% of the medium was renewed every other day.

For the growth experiments, ciliate cells were separated from their food source by repeated centrifugation (1600–1900×g, 3–5 min for *Colpidium*; 2000–2300×g, 3–5 min for *Cyclidium*) and resuspension in fresh medium, taking advantage of their negative geotaxis. Subsequently, the ciliate suspensions were slowly filtered through a 12-μm (*Colpidium*) or 5-μm (*Cyclidium*) membrane filter without vacuum, and cells retained on the filter were immediately resuspended in mineral water to obtain protist stock suspensions. Subsamples of the stock suspensions were taken to estimate the number of cells in the food suspensions using a Sedgewick-Rafter chamber. Cell sizes were determined by measuring the length and width of at least 50 unpreserved cells using an image analysis system (Tab. 1); the cell volume was computed using the geometric formula of a prolate spheroid (Hillebrand 1999). To estimate the contamination of the *Daphnia* food suspension with *Synechococcus*, subsamples were DAPI-stained and enumerated by epifluorescence microscopy. In the ciliate suspensions used to feed *Daphnia*, *Synechococcus* comprised <30 % of the total carbon. Ciliate suspensions used for analysis were prepared separately and more thoroughly, leading to a contamination of <15 % of the total carbon. Contamination with bacteria was negligible in all treatments (<0.6 % of total carbon; bacterial carbon was estimated according to Bratbak 1985).

Daphnia growth experiments—Growth experiments were conducted with third-clutch juveniles (born within 10 h) of a clone of *Daphnia magna*, originally isolated from Großer Binnensee, Germany (Lampert 1991). The experiments were carried out at 20°C in glass beakers filled with 200 ml of filtered lake water (0.45-μm pore-sized membrane filter). Each treatment consisted of three replicates with seven animals each. The food suspensions, containing 1.5 mg C l^{-1} of *Synechococcus* or *Cryptomonas* sp. (1.4×10^4 cells ml^{-1}) and at least 1 mg C l^{-1} (9×10^2 *Colpidium* ml^{-1}; 1×10^4 *Cyclidium* ml^{-1}) of ciliates, were renewed daily during the 6-day experiments. Somatic growth rates (g) were determined as the increase in dry weight (W) during the experiments using the equation:

$$g = \ln W_t - \ln W_0/t$$

Subsamples of the experimental animals consisting of seven individuals were taken at the beginning (W_0) and at the end (W_t) of an experiment. After 24 h in a drying chamber, subsamples were weighed on an electronic balance (Mettler UMT 2; $\pm$ 0.1 µg). Growth rates were calculated as means for each treatment.

Chemical analysis—For the analysis of fatty acids and neutral lipids (sterols and sterol-like compounds, e.g. tetrahymanol), 0.5 mg particulate organic carbon (POC) of the food suspensions was filtered on precombusted GF/F filters (Whatman, 25-mm diameter). Total lipids were extracted three times from filters with dichloromethane/methanol (2:1, v/v), and the pooled cell-free extracts were evaporated to dryness with nitrogen. The lipid extracts were transesterified with 3 mol methanolic HCl l^{-1} (60°C, 15 min) for the analysis of fatty acids or saponified with 0.2 mol methanolic KOH l^{-1} (70°C, 1 h) for the analysis of sterols. Subsequently, fatty acid methyl esters (FAMEs) were extracted three times with 2 ml of *iso*-hexane; neutral lipids were partitioned into *iso*-hexane:diethyl ether (9:1, v/v). The lipid-containing fraction was evaporated to dryness under nitrogen and resuspended in 10–20 µl *iso*-hexane.

Lipids were analyzed by gas chromatography on an HP 6890 GC equipped with a flame ionization detector and either a DB-225 (J&W Scientific) capillary column to analyze FAMEs or an HP-5 (Agilent) capillary column to analyze sterols. Details of GC configurations are given elsewhere (Von Elert 2002 for fatty acids, Martin-Creuzburg & Von Elert 2004 for sterols). Lipids were quantified by comparison to internal standards (C17:0 and C23:0 methyl esters; 5α-cholestan) and identified by their retention times and their mass spectra, which were recorded with a gas chromatograph/mass spectrometer (Finnigan MAT GCQ) equipped with a fused-silica capillary column (DB-225MS, J&W Scientific for FAMEs; DB-5MS, Agilent for sterols). Sterols and sterol-like compounds were analyzed in their free form and as their trimethylsilyl derivatives. Spectra were recorded between 50 and 600 amu in the EI ionization mode. Mass spectra were identified by comparison with mass spectra of reference substances (e.g. tetrahymanol and diplopterol, see acknowledgments) or spectra found in a self-generated spectra library or in the literature (e.g. Ten Haven et al. 1989, Venkatesan

1989, Harvey & McManus 1991). The detection limit was 20 ng of fatty acid or sterol. It was not possible to distinguish between petroselinic acid (C18:1n-12) and oleic acid (C18:1n-9). The absolute amount of each lipid was related to the POC. Therefore, aliquots of the food suspensions were filtered onto precombusted glass-fibre filters (Whatman GF/F, 25-mm diameter) and analysed for POC and nitrogen using an NCS-2500 analyzer (ThermoQuest GmbH, Egelsbach, Germany). For determination of particulate phosphorus, aliquots were collected on acid-rinsed polysulfon filters (HT-200; Pall, Ann Arbor, Mich., USA) and digested with a solution of 10 % potassium peroxodisulfate and 1.5 % sodium hydroxide for 60 min at 121°C, and soluble reactive phosphorus was determined using the molybdate-ascorbic acid method (Greenberg et al. 1985).

Data analysis—The somatic growth rates of *Daphnia magna* were analysed using one-way analysis of variance (ANOVA). Raw data met the assumption of homogeneity of variance (Levene's test); single treatment effects were tested by Tukey's HSD post-hoc test (p = 0.05).

Results

Microscopy revealed high grazing activities of both ciliate species on the two *Synechococcus* strains. The red cells of *Synechococcus* sp. strain BO8809 and the green cells of *S. elongatus* could easily be observed in the food vacuoles of the ciliates.

The two ciliate species differed significantly in size (Tab. 1), but cell sizes were not affected by the food source of the ciliates. The cell sizes of *Cryptomonas* sp. and *Cyclidium* were similar.

Growth experiments—Juvenile somatic growth rates (g) of *Daphnia magna* were significantly affected by the food supplied in the growth experiments (ANOVA, $F_{7,16} = 1146$; $p < 0.001$; Fig. 1). Growth rates of *D. magna* ranged from 0.01 d^{-1} without food supply to 0.51 d^{-1} when fed on *Cryptomonas* sp. The cyanobacteria *Synechococcus* sp. strain BO8809 and *S. elongatus* did not differ in their food quality for *D. magna* (Tukey's HSD, p = 1); growth on both strains was in general poor (g = 0.07 d^{-1}). In comparison to growth on pure *Synechococcus*, *D. magna* exhibited significantly higher growth rates on ciliates fed either of the two *Synechococcus* strains. Growth of *D. magna* was significantly higher on either of the ciliates fed on *S. elongatus* than on either of the ciliates fed on *Synechococcus* sp. strain

BO8809 (Tukey's HSD, p < 0.05). Regardless of the food source of the ciliates, *Cyclidium* improved the growth of *D. magna* more than *Colpidium* (Tukey's HSD, p < 0.05).

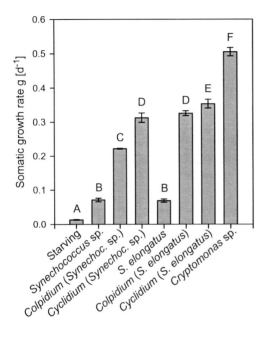

Fig. 1 Juvenile somatic growth rates of *Daphnia magna* grown on *Synechococcus* sp. strain BO8809, *Synechococcus elongatus*, or ciliates previously fed with one of the two *Synechococcus* strains. Growth without food supply (starving) and growth on *Cryptomonas* sp. are shown for comparison. Data are means of three replicates per treatment; error bars indicate SD. Bars labelled with the same letters are not significantly different (Tukey's HSD, p < 0.05 following ANOVA).

Table 1. Cell sizes of *Colpidium campylum* and *Cyclidium* sp. grown on one of two different *Synechococcus* strains, indicated in parentheses (*Synechococcus* sp. strain BO8809 or *Synechococcus elongatus*) and cell sizes of *Cryptomonas* sp. (n = 50; ± SD).

	Length [μm]			Width [μm]			Volume [μm³]
	Mean ± SD	Max.	Min.	Mean ± SD	Max.	Min.	Mean
Colpidium (*Synechococcus* sp.)	43.89 ± 3.91	52.33	36.20	19.97 ± 2.43	24.32	14.90	9164.74
Colpidium (*S. elongatus*)	44.41 ± 4.22	52.45	27.21	20.00 ± 2.65	25.82	12.12	9301.20
Cyclidium (*Synechococcus* sp.)	17.96 ± 2.05	22.25	12.41	8.87 ± 1.47	15.73	6.60	739.86
Cyclidium (*S. elongatus*)	17.81 ± 2.20	22.16	10.36	8.10 ± 1.08	11.29	5.63	611.83
Cryptomonas sp.	19.12 ± 1.68	23.09	15.01	8.82 ± 1.06	11.44	7.03	778.25

Elemental nutrient ratios—The carbon:nitrogen (C:N) and carbon:phosphorus (C:P) ratios of the various food suspensions are given in Table 2. The two *Synechococcus* strains were characterized by high nitrogen and phosphorus contents; however, the C:N and C:P ratios were slightly lower in *Synechococcus* sp. strain BO8809 than in *S. elongatus*. The C:N ratios of ciliates ranged from 4.40 to 4.92. The C:P ratios were in general lower in *Colpidium* than in *Cyclidium*. The nutrient ratios of the ciliates seemed to be affected by the nutrient ratios of their food source: the C:N and C:P ratios tended to be lower when the nitrogen- and phosphorus-rich *Synechococcus* sp. strain BO8809 was offered as food. The nitrogen content of *Cryptomonas* sp. was lower (higher C:N ratio) than that of the ciliates, whereas the C:P ratio was comparable to that of *Cyclidium*.

Table 2. Carbon:nitrogen (C:N) and carbon:phosphorus (C:P) molar ratios of the different food suspensions offered to *Daphnia magna* (n = 3; ± SD). The food source of the ciliates (*Synechococcus* sp. BO8809 or *S. elongatus*) is given in parentheses.

	Nutrient ratios	
	C:N	C:P
Synechococcus sp. strain BO8809	2.89 ± 0.62	79.87 ± 34.39
Synechococcus elongatus	4.46 ± 0.03	156.29 ± 12.13
Colpidium (*Synechococcus* sp.)	4.40 ± 0.16	58.78 ± 2.78
Colpidium (*S. elongatus*)	4.73 ± 0.03	71.79 ± 3.61
Cyclidium (*Synechococcus* sp.)	4.80 ± 0.08	113.19 ± 3.92
Cyclidium (*S. elongatus*)	4.92 ± 0.35	161.96 ± 3.73
Cryptomonas sp.	5.49 ± 0.07	123.95 ± 6.14

Sterols and sterol-like compounds—Sterols were not detected in the two *Synechococcus* strains or in the two ciliate species. Neutral lipids of *Colpidium* were characterized by the triterpenoid alcohol tetrahymanol (gammaceran-3β-ol) and its hopanoid isomer diplopterol. In *Cyclidium*, tetrahymanol was accompanied by hopan-3β-ol (Tab. 3). Stigmasterol (24-ethylcholesta-5,22-dien-3β-ol) and epibrassicasterol (24-methylcholesta-5,22-dien-3β-ol) were the principal sterols found in *Cryptomonas* sp. Epibrassicasterol, the 24α-epimer of brassicasterol, occurs in cryptophycean algae, such as *Cryptomonas* and *Rhodomonas* (Goad et al. 1983; Gladu et al. 1990). Although we did not determine the stereochemistry of the side chain at C-24, a 24α-configuration was also assumed.

Table 3. Neutral lipid content of *Colpidium campylum* and *Cyclidium* sp. (n = 3; ± SD; n.d. = not detected). The food source of the ciliates (*Synechococcus* sp. strain BO8809 or *Synechococcus elongatus*) is given in parentheses.

	Colpidium (*Synechococcus* sp.) [µg mg C^{-1}]	*Colpidium* (*S. elongatus*) [µg mg C^{-1}]	*Cyclidium* (*Synechococcus* sp.) [µg mg C^{-1}]	*Cyclidium* (*S. elongatus*) [µg mg C^{-1}]
Tetrahymanol	5.87 ± 1.64	8.41 ± 1.51	1.95 ± 0.31	2.73 ± 0.32
Diplopterol	10.48 ± 2.45	12.99 ± 2.39	n.d.	n.d.
Hopan-3β-ol	n.d.	n.d.	5.01 ± 0.17	5.28 ± 0.23
Total	16.35 ± 4.09	21.40 ± 3.90	6.96 ± 0.15	8.01 ± 0.51

Fatty acids—PUFAs were not detected in either *Synechococcus* strain, except for low amounts of 16:2n-6 in both strains and traces of 16:2n-4 in *Synechococcus* sp. strain BO8809 (Fig. 2). Instead, both strains were characterized by high amounts of short-chain saturated fatty acids and the monounsaturated fatty acid 16:1n-7. Compared to the fatty acid composition of the cyanobacteria, *Colpidium* contained high amounts of n-6 PUFAs, such as 16:2n-6, 18:2n-6, and 18:3n-6, and, in addition, moderate amounts of 16:2n-4, 16:3n-4, 18:1n-7, and 18:2n-7. In contrast, *Cyclidium* was characterized by high amounts of n-3 PUFAs (18:3n-3, 18:4n-3) rather than n-6 PUFAs (Fig. 2). Dietary effects on the fatty acid content of the ciliates were negligible. The fatty acid composition of *Cryptomonas* sp. was dominated by high amounts of n-3 PUFAs, such as 18:3n-3, 18:4n-3, and 20:5n-3. The total fatty acid content was highest in *Cryptomonas* sp. (211.0 ± 12.6 µg mg C^{-1}) and lowest in the *Synechococcus* strains (79.3 ± 4.9 µg mg C^{-1} in *Synechococcus* sp. strain BO8809; 98.8 ± 4.6 µg mg C^{-1} in *S. elongatus*). The ciliates showed intermediate contents of total fatty acids: 121.7 ± 2.3 µg mg C^{-1} in *Colpidium* fed on *Synechococcus* sp. strain BO8809; 112.0 ± 4.7 µg mg C^{-1} in *Colpidium* fed on *S. elongatus*; 131.7 ± 13.0 µg mg C^{-1} in *Cyclidium* fed on *Synechococcus* sp. strain BO8809; and 161.4 ± 18.4 µg mg C^{-1} in *Cyclidium* fed on *S. elongatus*.

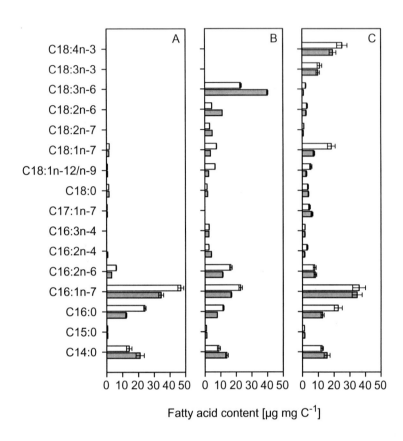

Fig. 2 Fatty acid content of *Synechococcus* sp. strain BO8809 (A, grey bars), *Synechococcus elongatus* (A, white bars), and *Colpidium campylum* (B) and *Cyclidium* sp. (C) previously fed on either *Synechococcus* sp. strain BO8809 (grey bars) or *S. elongatus* (white bars). Data are means of three replicates; error bars indicate SD.

Discussion

In oligotrophic to mesotrophic lakes, picocyanobacteria, such as *Synechococcus* species, are abundant autotrophic prokaryotes that often contribute significantly to primary production, thereby forming the base of a complex pelagic food web (Weisse 1993, Callieri & Stockner 2002). Grazing by ciliates can be considered as an important loss process controlling biomass and production of APP in freshwater ecosystems (Fahnenstiel et al. 1991, Sherr et al. 1991, Šimek et al. 1995, Hadas & Berman 1998).

The two ciliate species (*Colpidium campylum*, *Cyclidium* sp.) used in the present study showed high grazing activities on both *Synechococcus* strains. Although bacterial contamination was marginal, the consumption of bacteria could not be excluded since none of the ciliate cultures was axenic. However, it has already been noted that various ciliate species (e.g. *Cyclidium* sp.) can meet all carbon requirements exclusively from an APP diet (Šimek et al. 1996), and, moreover, Šimek et al. (1996) reported a strong size selection towards larger picoplankton prey; therefore, the consumption of bacteria in the present study appears negligible.

Invertebrate grazers, in particular cladocerans of the genus *Daphnia*, feed on a wide size-range of particles that includes picocyanobacteria and small ciliates (Jürgens 1994). It is well established that picocyanobacteria, such as *Synechococcus*, are unsuitable for supporting growth of *Daphnia* (Lampert 1977*a,b*, DeMott & Müller-Navarra 1997, Von Elert et al. 2003, Martin-Creuzburg & von Elert 2004, Martin-Creuzburg et al. 2005), whereas effects of ciliates on *Daphnia* nutrition are controversial. In the field, ciliates are suppressed during population peaks of *Daphnia*, and several experimental studies have shown that daphnids effectively prey on ciliates up to a certain size (Tezuka 1974, Porter et al. 1979, Jack & Gilbert 1993, Sanders et al. 1996). However, ciliates are less nutritious for *Daphnia* than many algae (DeBiase et al. 1990, Wickham et al. 1993, Sanders et al 1996). In the present study, we addressed whether ciliates can serve as a trophic link between APP production and *Daphnia* production, thereby upgrading the nutritional value of an APP food source. Both *Synechococcus* strains used here were unsuitable as a sole food source for *D. magna*, whereas either of the two ciliate species as a food source, which fed on the *Synechococcus* strains themselves, significantly enhanced the somatic growth of *D. magna* . The results obtained from this simplified experimental food chain (APP→ciliates→*Daphnia*) suggest that predation on ciliates by *Daphnia* allows access to APP production and provides a linkage of carbon flow to higher trophic levels. Moreover, ciliates obviously improve the quality of the supplied food by the addition of essential components that are absent in *Synechococcus*.

S. elongatus is a nontoxic picocyanobacterium that is well-assimilated by *Daphnia* (Lampert 1977*a,b*, 1981). The determined elemental nutrient ratios showed that both *Synechococcus* strains and, possibly as a consequence, both ciliate species were rich in phosphorus and nitrogen, which is in accordance with previous data (DeBiase et al. 1990, Sanders et al. 1996, DeMott 1998). In all treatments, C:P ratios were far below the C:P ratios determined for P-limited growth of *Daphnia* (C:P > 300; Sterner & Schulz 1998), and C:N ratios were considerably lower than C:N ratios usually found in phytoplankton species (e.g. *Cryptomonas* sp. in the present study). Thus, the growth-enhancing trophic upgrading effect is unlikely caused by the elemental nutrient supply.

Stoecker & McDowell Capuzzo (1990) proposed that ciliates might be an important source of essential lipids, thus representing a supplementary diet to enhance growth of *Daphnia*. The lack of sterols in *Synechococcus* has recently been identified as the major food quality constraint in *Daphnia* (Von Elert et al. 2003, Martin-Creuzburg et al. 2005), which implies that the observed trophic upgrading of APP food quality by intermediary ciliates is due to the addition of sterols. However, available data suggest that ciliates lack the ability to synthesize sterols de novo (Conner 1968, Harvey & McManus 1991, Harvey et al. 1997, Klein Breteler et al. 2004). Instead, the pentacyclic triterpenoid alcohol tetrahymanol is produced, which is functionally equivalent to sterols as a structural component of cell membranes. Numerous studies have shown that exogenously supplied sterols can be incorporated into cell membranes of ciliates, and as a consequence thereof, inhibit tetrahymanol synthesis (first demonstrated by Conner et al. 1968). Hence, the occurrence of sterols in ciliates depends on their diet: ciliates feeding on bacteria or picocyanobacteria cannot rely on a dietary source of sterols and, therefore, are expected to produce tetrahymanol. Sterols were not detected in either of the two ciliate species used in the present study; this lack of sterols can be attributed to the absence of sterols in their food source, *Synechococcus*. Instead, relatively high amounts of tetrahymanol and its isomer diplopterol were detected in *Colpidium*; *Cyclidium* contained considerably lower amounts of tetrahymanol and no diplopterol. Tetrahymanol was first identified in *Tetrahymena pyriformis* (Mallory et al.1963), a ciliate related to *Colpidium*, but has recently also been observed in several marine scuticociliates, including *Cyclidium* sp. (Harvey & McManus 1991). Neutral lipids of *Cyclidium* were dominated by hopan-3β-ol, a hopanoid that seems to be common among scuticociliates (Harvey & McManus 1991, Harvey et al. 1997).

Like all arthropods, crustaceans are incapable of synthesizing sterols de novo and therefore must acquire these essential nutrients from their diet (Goad 1981). Daphnids feeding on bacterivorous or picocyanobacterivorous ciliates cannot rely on a dietary source of sterols and, therefore, are expected to be sterol limited. However, tetrahymanol and related compounds that functionally replace sterols as membrane reinforcers in ciliates (Raederstorff & Rohmer 1988) might also be suitable as sterol surrogates in crustacean tissues. Thus, the observed trophic upgrading of *Synechococcus* food quality by intermediary ciliates might be due to the addition of tetrahymanol and hopanoids, which at least partly release *Daphnia* from sterol limitation. Ederington et al. (1995) reported the assimilation of tetrahymanol in tissues of ciliate-fed copepods (mainly in eggs) and suggested that tetrahymanol can provide functional equivalence to cholesterol, thereby maintaining minimal egg production. Beside their role as structural components of cell membranes, sterols serve as precursors for many bioactive molecules, such as ecdysteroids, which are involved in the process of molting (Goad 1981). Whether tetrahymanol and related compounds affect growth and/or reproduction of crustaceans remains to be tested, possibly by supplementation of these compounds to a sterol-free diet. Recently, Klein Breteler et al. (2004) found no evidence for trophic upgrading of a sterol-deficient diet by the marine ciliate *Strombidium sulcatum* for copepods. However, neither sterols nor tetrahymanol nor other sterol surrogates were detected in *S. sulcatum*, which possibly makes this bacterivorous ciliate unsuitable as a single food source for copepods, and corroborates the finding of this study that trophic upgrading of a sterol-free diet can be attributed to tetrahymanol-related compounds in ciliates.

Von Elert et al. (2003) have shown that when the shortage of sterols in *Synechococcus* is overcome by supplementation with cholesterol, growth of *Daphnia galeata* is limited by the availability of long-chain PUFAs. Hence, the addition of PUFAs by the intermediary ciliates might have further improved the trophic upgrading effect, provided that the daphnids were released from sterol limitation.

Long-chain PUFAs were not detected in either of the *Synechococcus* strains. However, the fatty acid composition of *Colpidium* was characterized by relatively high amounts of n-6 PUFAs (18:2n-6, 18:3n-6), which suggested a de-novo synthesis of these fatty acids by *Colpidium*. This is corroborated by the finding that species of *Tetrahymena* synthesize n-6 PUFAs when grown on fatty-acid-free media (Sul & Erwin 1997). In contrast, *Cyclidium* was found to synthesize high amounts of n-3 PUFAs (18:3n-3, 18:4n-3), which previously have been detected in marine scuticociliates: 18:3n-3 in *Pleuronema* sp. (Ederington et al. 1995), and 18:3n-3 and 18:4n-3 in *Parauronema acutum* (Sul & Erwin 1997, Sul et al. 2000).

In most animals, the PUFAs 18:2n-6 and 18:3n-3 are essential dietary compounds that play important roles in animal physiology (Cook 1996). In *Daphnia*, the n-6 PUFAs found in *Colpidium* (18:2n-6, 18:3n-6) might be further converted into 20:4n-6, an intermediate in prostaglandin synthesis (Weers et al. 1997). However, laboratory experiments and correlative field studies suggest either 18:3n-3 or 20:5n-3 as a potentially limiting resource that constrains proper growth of *Daphnia* (Müller-Navarra 1995, Wacker & Von Elert 2001, Müller-Navarra et al. 2000, Von Elert 2002, Becker & Boersma 2003, Ravet et al. 2003). In

Cyclidium, comparatively high amounts of n-3 PUFA (18:3n-3, 18:4n-3) were detected, which, in *Daphnia*, can be converted into 20:5n-3 through a process of elongation and desaturation. Although the conversion of 18:3n-3 into 20:5n-3 is low (Weers et al. 1997, Von Elert 2002), the availability of these PUFAs might be adequate to meet metabolic demands. The finding that *Cyclidium*, even though it contains less tetrahymanol and related compounds than *Colpidium*, resulted in a higher trophic upgrading of APP, leads to the conclusion that the amount of tetrahymanol-related compounds present in *Cyclidium* was sufficient to release daphnids from the sterol limitation observed on pure *Synechococcus* (Martin-Creuzburg et al. 2005; Von Elert et al. 2003). Hence the superior quality of *Cyclidium* must be attributed to compounds present in *Cyclidium* but not in *Colpidium*. In accordance with the finding that when sterol requirements are met, growth of daphnids on *Synechococcus* becomes limited by n-3 PUFAs (Von Elert et al. 2003), the synthesis of n-3 PUFAs in *Cyclidium* but not in *Colpidium* provides a reasonable explanation for the superior food quality of *Cyclidium* in comparison to *Colpidium*.

Differences in the food quality of *Colpidium* and *Cyclidium* for *D. magna* might also be due to their different cell sizes. Jack & Gilbert (1993) reported that large ciliates are less susceptible to *Daphnia* predation and showed that clearance rates decrease with increasing ciliate size. *Colpidium* might be close to the upper size range of food particles that can be ingested as a whole by juvenile *D. magna*. Large particles up to a certain size have to be manipulated, which results in a reduced ingestion efficiency (Porter et al. 1979). However, ciliates much larger than *Colpidium* have been shown to be ingestible by *Daphnia pulex*, and among various ciliates tested, a ciliate similar in size to *Colpidium campylum* (*Tetrahymena pyriformis*) was most vulnerable to *Daphnia* predation (Jack & Gilbert 1993). Therefore, it seems unlikely that the observed differences in food quality of the two ciliates are exclusively derived from differences in cell sizes.

Although the ciliates used here improved APP food quality for *Daphnia*, growth rates achieved on a ciliate diet rank far below those achieved with *Cryptomonas* sp., which contained comparatively high amounts of the Δ^5 sterols stigmasterol and epibrassicasterol. Δ^5 sterols such as stigmasterol have been shown to support somatic growth and reproduction of daphnids when supplemented to the sterol- and PUFA-deficient *S. elongatus* (Martin-Creuzburg & Von Elert 2004). In addition, the fatty acid composition of *Cryptomonas* sp. was dominated by high levels of n-3 PUFAs, especially 18:3n-3, 18:4n-3, and 20:5n-3, which is in accordance with previous data (e.g. Von Elert & Stampfl 2000). This implies that the high food quality of *Cryptomonas* sp. is a combined effect of its sterol and PUFA composition.

In summary, the presented data clearly show that predation on ciliates by *Daphnia* can provide a linkage between APP production and zooplankton production. Especially in oligotrophic to mesotrophic lakes, where APP species often dominate phytoplankton assemblages, this might be an important pathway channelling carbon and essential nutrients to higher trophic levels. Daphnids have been shown to be primarily sterol-limited when grown on the picocyanobacterium *Synechococcus*, which implies that the observed trophic upgrading of APP food quality by intermediary ciliates is due to the addition of sterols or sterol-like compounds that at least partly release *Daphnia* from sterol limitation. The absence of sterols in ciliates suggests that tetrahymanol and/or hopanoids are functional equivalents to sterols not only in ciliates but also in *Daphnia*, thereby leading to enhanced growth.

Acknowledgments—We thank P. Merkel and C. Gebauer for excellent technical assistance and M. Rohmer for providing us with standards of tetrahymanol and diplopterol. This work was supported by the German Research Foundation (DFG, EI 179/4-2), and by a post-doc fellowship to A. B. from the Conseil Régional d'Auvergne, France.

Chapter 6

Food quality of ciliates for *Daphnia*: the role of sterols

Dominik Martin-Creuzburg, Alexandre Bec and Eric von Elert

Abstract—Experimental results provide evidence that trophic interactions between ciliates and *Daphnia* are constrained by the comparatively low food quality of ciliates. The dietary sterol content is a crucial parameter in determining food quality for *Daphnia*. Ciliates, however, do not synthesize sterols de-novo. We hypothesized that ciliates are nutritionally inadequate because of their lack of sterols and tested this hypothesis in growth experiments with *Daphnia magna* and the ciliate *Colpidium campylum*. The lipid content of the ciliate was altered by feeding on fluorescently labeled albumin beads supplemented with different sterols. Ciliates that preyed upon a sterol-free diet (bacteria) did not contain any sterols, and growth of *D. magna* on these ciliates was poor. Supplementation of the ciliates' food source with different sterols led to the incorporation of the supplemented sterols into the ciliates' cell membranes and to enhanced growth of *D. magna*. Sterol limitation was thereby identified as the major constraint of ciliate food quality for *Daphnia*. Furthermore, by supplementation of sterols unsuitable for supporting *Daphnia* growth, we provide evidence that ciliates as intermediary grazers biochemically upgrade unsuitable dietary sterols to sterols appropriate to meet the physiological demands of *Daphnia*.

Keywords—*Colpidium*, *Daphnia magna*, dietary sterols, tetrahymanol

Reports on trophic interactions between ciliates and *Daphnia* are controversial. Field experiments have indicated that ciliates are suppressed during population peaks of *Daphnia* by facing a substantial grazing pressure (Carrick et al. 1991; Pace and Funke 1991; Wickham and Gilbert 1991; Marchessault and Mazumder 1997; Jürgens et al. 1999; Zöllner et al. 2003), and several laboratory studies have revealed that cladocerans effectively prey on ciliates up to a certain size, which implies that daphnids are important ciliate predators (Tezuka 1974; Porter et al. 1979; Jack and Gilbert 1993; Sanders et al. 1996). However, although rich in nitrogen and phosphorus, ciliates are less nutritious for daphnids than many algae (DeBiase et al. 1990; Wickham et al. 1993; Sanders et al 1996; Bec et al. 2003).

Recently, the dietary sterol content has been identified as a crucial parameter in determining food quality for *Daphnia* (Von Elert et al. 2003; Martin-Creuzburg and Von Elert 2004; Martin-Creuzburg et al. 2005). Available data suggest that ciliates lack the ability to synthesize sterols de-novo (e.g., Conner et al. 1968; Harvey and McManus 1991; Harvey et al. 1997; Klein-Breteler et al. 2004), which led us to hypothesize that the observed nutritional inadequacy of ciliates is caused by their lack of sterols. This hypothesis is complicated by the finding that ciliates are able to incorporate exogenously supplied sterols into cell membranes and to metabolize them to various sterols (Conner et al. 1968; Harvey and McManus 1991; Harvey et al. 1997). In the absence of exogenous sterols, ciliates produce the pentacyclic triterpenoid alcohol tetrahymanol, which is functionally equivalent to sterols as a structural component of cell membranes (Conner et al. 1968). Hence, the occurrence of sterols in ciliates, and thereby the food quality of the ciliates for higher trophic levels, might depend on the food source preyed upon by the ciliates, i.e., ciliates feeding on a sterol-free diet (most prokaryotes, Volkman 2003) are expected to produce tetrahymanol, and ciliates feeding on eukaryotic food sources are expected to incorporate dietary sterols, a process that might be associated with changes in the structure of the sterols via metabolism. Since sterols differ in their suitability to support *Daphnia* growth (Martin-Creuzburg and Von Elert 2004), these structural changes might also affect the food quality of ciliates for *Daphnia*.

Here we assessed the importance of sterols in determining the food quality of ciliates for *Daphnia*. *Daphnia magna* was reared on the ciliate *Colpidium campylum,* whose lipid content was altered by feeding fluorescently labeled albumin beads supplemented with different sterols.

Cultivation and preparation of the food organisms—Colpidium campylum was obtained from the Laboratoire de Biologie des Protistes (Université Blaise Pascal, France). It was cultivated semi-continuously at 20°C in mineral water (Volvic®) on an undefined bacterial assemblage. Exponential growth of the ciliates was maintained by exchanging 20 to 40% of the medium every other day. Autoclaved wheat grains were added to enhance bacterial growth. Ciliate food suspensions for the *Daphnia magna* growth experiments were prepared by feeding *C. campylum* fluorescently labeled albumin beads (2 μm, Micromod, Rostock, Germany) supplemented with cholesterol, lathosterol, or dihydrocholesterol (Sigma). For the supplementation, 100 μl of a bead stock solution (2.4×10^9 beads ml^{-1} ethanol), which was previously sonicated for 10 min in an ultrasonic bath to eliminate clumping, was dissolved in 10 ml ethanol. Subsequently, 100 μl of a sterol stock solution (2.5 mg ml $ethanol^{-1}$) was added and incubated with beads for 30 min at 20°C. The bead/sterol suspensions were evaporated to dryness under a stream of nitrogen. The loaded beads were resuspended in 10 ml mineral water and added to 200 ml of a ciliate suspension at a concentration of approximately 1500 cells ml^{-1}. Each *C. campylum* suspension was fed loaded beads 4 and 2 days before the ciliates were fed to *D. magna*.

For the growth experiments, ciliate cells were separated from bacteria and albumin beads by repeated centrifugation ($1600–1900 \times g$; 4 min) and resuspension in fresh medium, taking advantage of the negative geotaxis of the ciliates. Subsequently, the ciliate suspensions were slowly filtered through a 12-μm membrane filter without vacuum, and retained cells on the filter were immediately resuspended in mineral water to obtain the protist food suspensions used in the *D. magna* growth experiments. Subsamples were taken to estimate the number of cells in the food suspensions using a Sedgewick-Rafter chamber; the carbon concentrations were determined with an NCS-2500 analyzer (ThermoQuest GmbH, Egelsbach, Germany). To estimate the contamination of the food suspension with bacteria and albumin beads, subsamples were stained with DAPI, and cells were enumerated by epifluorescence microscopy. Contamination of the food suspensions with bacteria was negligible in all treatments (<0.83% of total carbon; bacterial carbon was estimated according to Bratbak 1985). Free albumin beads were only occasionally observed in the ciliate food suspensions used to feed *D. magna*, and never exceeded a concentration of 2.5×10^2 beads ml^{-1}.

The green alga *Scenedesmus obliquus* (SAG 276-3a), a fairly good food source for *Daphnia* (Martin-Creuzburg et al. 2005), was used as food for the stock cultures of *D. magna* and as a reference food in the growth experiments. Culture conditions and preparation of algal stock solutions are described elsewhere (Martin-Creuzburg et al. 2005).

Daphnia growth experiments—Third clutch juveniles (born within 8 h) of a clone of *D. magna* (isolated by Lampert 1991) were used in growth experiments. The experiments were carried out at 20°C in glass beakers filled with 100 ml of filtered lake water (0.45-μm pore sized). Each treatment consisted of three replicates with four animals each. The food suspensions contained 1.5 mg C L^{-1} of *S. obliquus* and at least 1 mg C L^{-1} (~900 cells ml^{-1}) of *C. campylum* and were renewed daily within the 6-day experiments. Juvenile somatic growth rates were determined according to Martin-Creuzburg et al. (2005).

Growth rates of *D. magna* were significantly affected by the food supplied in the growth experiment (ANOVA, $F_{5,12} = 258.6$; $p < 0.001$; Fig. 1). Growth of *D. magna* on *C. campylum* did not differ from growth on *C. campylum* previously fed with unsupplemented albumin beads (Tukey's HSD, $p = 0.22$). Supplementation of the albumin beads with any one of the three sterols prior to feeding to *C. campylum* improved the growth of *D. magna* (Tukey's HSD, $p < 0.05$). Supplementation with lathosterol had a lower ameliorating effect on growth than supplementation with cholesterol or dihydrocholesterol.

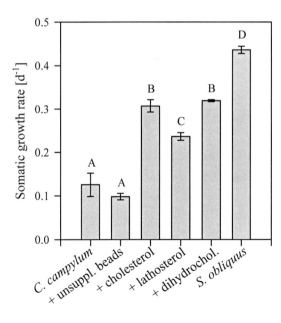

Fig. 1 Juvenile somatic growth rates of *Daphnia magna* grown on *Colpidium campylum* previously fed with bacteria (bar labeled with *C. campylum*), bacteria and unsupplemented albumin beads, or bacteria and albumin beads supplemented with cholesterol, lathosterol, or dihydrocholesterol. Growth of *D. magna* on *Scenedesmus obliquus* is given as a reference. Data are means of three replicates per treatment; error bars indicate SD.

Ciliate uptake of fluorescently labeled albumin beads—The uptake of sterol-supplemented albumin beads by *C. campylum* was assessed by incubating ciliate cells (1500 cells ml^{-1}) for 20 min with loaded beads (1.2×10^6 beads ml^{-1}) at 20°C without mixing. Preliminary experiments revealed an incubation time of 20 min as the best compromise between bead uptake, countability in the food vacuoles, and digestion-egestion processes. Ingestion was stopped by adding ice-cold glutaraldehyde (final concentration 2%). Subsamples were stained with DAPI (1 $\mu g\ ml^{-1}$, 8 min) and gently filtered onto 0.2-μm black nucleopore filters. The uptake of the fluorescent beads by ciliates was determined using epifluorescence microscopy. In each replicate, beads were counted in at least 50 ciliate cells at $\times 1000$ magnification.

The ingestions rates of albumin beads by *C. campylum* and the proportions of labeled cells were not affected by supplementation with sterols (ANOVA, $F_{3,8} = 2.14$; $p = 0.17$ and $F_{3,8} = 3.46$; $p = 0.07$ respectively; Tab. 1). In each treatment, approximately 50–60% of the ciliate cells were fluorescently labeled. Unattached sterols can also be incorporated directly from the culture medium (e.g., Nes et al. 1981; Harvey et al. 1997); this could have further increased the sterol loading of the ciliate cells.

Table 1. Ingestion of unsupplemented or sterol-supplemented albumin beads by *Colpidium campylum* (values are means of $n = 3$; ± SD).

	Number [beads $cell^{-1}$]	Range [beads $cell^{-1}$]	Labeled cells [%]	Ingestion rate [beads $h^{-1}\ cell^{-1}$]
C. campylum + unsupplemented beads	0.99 ± 0.13	0–5	61.11 ± 2.55	2.98 ± 0.40
C. campylum + cholesterol-supplemented beads	1.06 ± 0.22	0–4	63.89 ± 6.94	3.19 ± 0.65
C. campylum + lathosterol-supplemented beads	0.77 ± 0.14	0–4	51.67 ± 4.41	2.32 ± 0.41
C. campylum + dihydrocholesterol-supplemented beads	1.12 ± 0.22	0–6	60.56 ± 4.81	3.37 ± 0.66

Results obtained from experiments with the ciliate *Tetrahymena pyriformis* suggest that the incorporation of supplemented sterols into cell membranes results in a reduced volume of the ciliate cells, possibly because of an increased ordering of the membrane components (Conner et al. 1982). According to Jack and Gilbert (1993), larger ciliates are less susceptible to *Daphnia* predation, and clearance rates increase with decreasing ciliate size; hence, a reduced cell volume of *C. campylum* might facilitate the ingestion process. Therefore, the mean cell size in each replicate was determined by measuring at least 50 unpreserved cells in length and width using an image analysis system, and the cell volume was computed using the geometric formula of a prolate spheroid (Tab. 2). Although the incorporation of supplemented sterols into ciliate cell membranes slightly reduced the cell volume of the protists, this effect was not significant (Tukey's HSD, $p > 0.05$) and therefore considered to be negligible. It should be mentioned that *Daphnia* can ingest ciliates much larger than *C. campylum*, and among various ciliates tested, *T. pyriformis,* a ciliate similar in size to *C. campylum,* was most vulnerable to *Daphnia* predation (Jack and Gilbert 1993).

Table 2. Size of *Colpidium campylum* grown on bacteria, bacteria and albumin beads, or bacteria and albumin beads supplemented with cholesterol, lathosterol, or dihydrocholesterol (values are means of $n = 3$; ± SD).

	Length [μm]	Width [μm]	Volume [μm³]
C. campylum	41.84 ± 0.42	19.67 ± 1.63	8515 ± 1411
C. campylum + unsupplemented beads	40.60 ± 0.71	19.00 ± 0.49	7683 ± 477
C. campylum + cholesterol-supplemented beads	38.87 ± 0.65	17.88 ± 1.48	6544 ± 1138
C. campylum + lathosterol-supplemented beads	39.78 ± 1.60	17.11 ± 1.33	6842 ± 722
C. campylum + dihydrocholesterol-supplemented beads	38.17 ± 0.74	17.45 ± 0.46	6088 ± 332

Lipid analyses—Lipids were extracted three times from pre-combusted GF/F filters (Whatman, 25-mm diameter) loaded with approximately 0.5 mg particulate organic carbon (POC) of the protist food suspensions using a mixture of dichloromethane/methanol (2:1, v/v). For the analysis of neutral lipids (sterols, tetrahymanol, and diplopterol), the pooled cell-free extracts were dried under a stream of nitrogen and saponified with 0.2 mol L^{-1} methanolic KOH (70°C, 1 h). Subsequently, neutral lipids were partitioned into *iso*-hexane:diethyl ether (9:1, v/v), dried under a stream of nitrogen, and resuspended in a volume of 10–20 μl *iso*-hexane. Neutral lipids were analyzed with a gas chromatograph (HP 6890, Agilent Technologies, Waldbronn, Germany) equipped with an HP-5 capillary column (Agilent) and a flame ionization detector. Details of the GC configurations are given elsewhere (Martin-Creuzburg and Von Elert 2004). Lipids were quantified as cholesterol equivalents by comparison to an internal standard (5α-cholestan). The detection limit was approximately 20 ng of sterol. Lipids were identified by their retention times and their mass spectra, which were recorded with a gas chromatograph/mass spectrometer (Finnigan MAT GCQ) equipped with a fused-silica capillary column (DB-5MS, Agilent). Sterols were analyzed in their free form and as their acetate and trimethylsilyl derivatives. Mass spectra were recorded between 50 and 600 amu in the EI ionization mode, and lipids were identified by comparison with mass spectra of reference substances purchased from Sigma or Steraloids (see also the Acknowledgments) and/or mass spectra found in the literature (e.g., Ten Haven et al. 1989; Venkatesan 1989; Harvey and McManus 1991). The absolute amount of each lipid was related to the POC, which was determined from another aliquot of the food suspensions used for lipid analysis.

Several sterols were detected in the wheat grains (dominated by sitosterol, campesterol and its stanols, which are commonly found in wheat grains, e.g. Takatsuto et al. 1999) used to enhance bacterial growth, but no sterols were detected in the bacterial fraction of the ciliate cultures. *C. campylum* cells grown without sterol-supplemented beads did not contain any polycyclic alcohols except tetrahymanol (gammaceran-3β-ol) and diplopterol (hopan-22-ol), which suggests that a dietary source of sterols for *C. campylum* was not available.

Sterols provided by the supplemented beads were incorporated and modified by *C. campylum* (Tab. 3). Moreover, the incorporation of supplemented sterols into cell membranes reduced the tetrahymanol and diplopterol content of the ciliate cells (Tukey's HSD, $p < 0.05$; ANOVA, $F_{4,10} = 31.61$ and $F_{4,10} = 40.59$; $p < 0.001$; Fig. 2). The inhibition of tetrahymanol production by exogenously supplied sterols has been extensively studied in *T. pyriformis*, an experimental model ciliate related to *C. campylum*. Thereby, cholesterol, lathosterol, dihydrocholesterol, and several other sterols have been shown to inhibit tetrahymanol and diplopterol production (e.g., Conner et al. 1969; Conner et al. 1978).

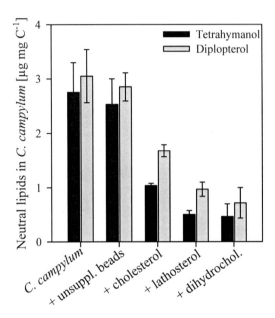

Fig. 2 Tetrahymanol and diplopterol content of *Colpidium campylum* grown on bacteria (bar labeled with *C. campylum*), bacteria and unsupplemented albumin beads, or bacteria and albumin beads supplemented with cholesterol, lathosterol, or dihydrocholesterol. Data are means of three replicates per treatment; error bars indicate SD.

Supplemented cholesterol was accumulated in high amounts in *C. campylum* cells and metabolized to cholesta-5,22-dien-3β-ol and cholesta-5,7,22-trien-3β-ol (Tab. 3). The desaturation of cholesterol to cholesta-5,7,22-trien-3β-ol via cholesta-5,22-dien-3β-ol has been studied in detail in *T. pyriformis* (e.g., Conner et al. 1969). Sterols containing double bonds at Δ^5, $\Delta^{5,22}$, or $\Delta^{5,7,22}$ enhance the growth of *Daphnia* on a sterol-free diet (Martin-Creuzburg and von Elert 2004). Therefore, the improved growth of *D. magna* on cholesterol-supplemented *C. campylum* (Fig. 1) can be attributed to the accumulation of these sterols in the ciliate cells.

The accumulation of supplemented lathosterol in *C. campylum* was comparatively low (Tab. 3). In *C. campylum*, lathosterol was metabolized to cholesta-7,22-dien-3β-ol and finally to cholesta-5,7,22-trien-3β-ol, the same end product found for cholesterol metabolism. Lathosterol is also efficiently converted to cholesta-5,7,22-trien-3β-ol in *T. pyriformis*

(Mulheirn et al. 1971). Growth of *D. magna* on lathosterol-supplemented *C. campylum* was enhanced, but ranked below the growth on cholesterol-supplemented ciliates (Fig. 1). The Δ^7-sterol lathosterol has been previously shown to support *Daphnia* growth to a significantly lower extent than Δ^5-sterols (Martin-Creuzburg and von Elert 2004). Therefore, the reduced growth of *D. magna* on lathosterol-supplemented *C. campylum* might be due to (i) the higher content of Δ^7-sterols, which are less efficiently used by daphnids, or (ii) to an overall lower sterol content of the ciliates that is insufficient in releasing *Daphnia* from sterol limitation. Klein Breteler et al. (1999) have suggested that the heterotrophic dinoflagellate *Oxyrrhis marina*, as an intermediary grazer, might upgrade a poor-quality food source containing only Δ^7-sterols, which are not suitable for supporting the development of copepods, by producing Δ^5-sterols. Even if ciliates do not synthesize sterols de-novo, they might upgrade the biochemically mediated food quality for higher trophic levels by accumulating and modifying dietary sterols, e.g. by converting Δ^7-sterols to $\Delta^{5,7,22}$-sterols, which are highly efficient in supporting *Daphnia* growth (Martin-Creuzburg and Von Elert 2004).

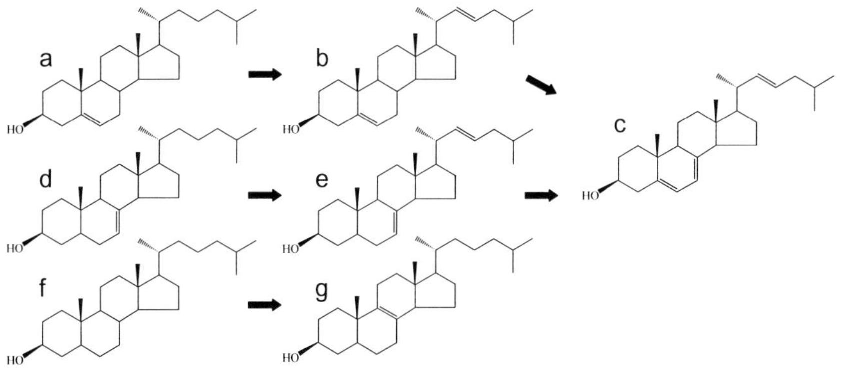

Fig. 3 Dehydrogenation of supplemented sterols by *Colpidium campylum*: cholest-5-en-3β-ol (cholesterol, a) possibly via cholesta-5,22-dien-3β-ol (b) and cholest-7-en-3β-ol (lathosterol, d) possibly via cholesta-7,22-dien-3β-ol (e) to cholesta-5,7,22-trien-3β-ol (c); cholestan-3β-ol (dihydrocholesterol, f) to cholest-8-en-3β-ol (g, tentative identification).

High amounts of supplemented dihydrocholesterol were found in the ciliate cells, accompanied by even higher amounts of a sterol tentatively identified as cholest-8-en-3β-ol. This suggested an effective conversion by the ciliate of the completely saturated dihydrocholesterol to sterols unsaturated at Δ^8 in the sterol nucleus. Although lanosterol, a

$\Delta^{8,24}$-sterol with additional C-4 dimethyl and C-14 methyl substituents, did not improve the growth of *Daphnia* when supplemented to a sterol-free diet (Martin-Creuzburg and Von Elert 2004), we propose that the Δ^8-sterol produced by the ciliate enhanced *Daphnia* growth. Differences in the suitability for *Daphnia* of lanosterol and the Δ^8-sterol found in the present study might be attributed to the additional methyl groups present in lanosterol, which affect important properties of the sterol molecule (Yeagle et al. 1977; Nes et al. 1978). The conversion of dietary Δ^8-sterols to cholesterol in copepods has already been proposed by Harvey et al. (1987). In contrast to *C. campylum*, *T. pyriformis* is capable of introducing a Δ^5, Δ^7, and Δ^{22} double bond into dihydrocholesterol, leading to the formation of cholesta-5,7,22-trien-3β-ol (Mulheirn et al. 1971). In both cases, however, this leads to an upgrading of food quality by the ciliate since dihydrocholesterol itself does not support the growth of *Daphnia* (Martin-Creuzburg and Von Elert 2004).

Table 3. Sterol content of *Colpidium campylum* grown on bacteria and albumin beads supplemented with cholesterol, lathosterol, or dihydrocholesterol (values are means of $n = 3$; ± SD).

	Sterol	Sterol concentration [$\mu g \ mg \ C^{-1}$]
C. campylum + cholesterol- supplemented beads	Cholest-5-en-3β-ol (cholesterol)	10.54 ± 3.27
	Cholesta-5,22-dien-3β-ol	18.30 ± 4.73
	Cholesta-5,7,22-trien-3β-ol	trace[a]
	Total	28.84 ± 7.99
C. campylum + lathosterol- supplemented beads	Cholest-7-en-3β-ol (lathosterol)	3.08 ± 1.54
	Cholesta-7,22-dien-3β-ol	1.05 ± 0.38
	Cholesta-5,7,22-trien-3β-ol	2.66 ± 1.68
	Total	6.79 ± 3.59
C. campylum + dihydrocholesterol- supplemented beads	Cholestan-3β-ol (dihydrocholesterol)	12.83 ± 6.53
	Cholest-8-en-3β-ol[b]	17.91 ± 4.91
	Total	30.74 ± 11.44

[a] Potentially underestimated because the retention time overlapped with that of cholesterol (cp. Valcarce et al. 2000).
[b] Tentative identification.

A number of physiological effects of sterol incorporation into ciliate cell membranes have also been reported, e.g., a reduction in cellular volume (as described above) and an elevation in saturated fatty acids (Ferguson et al. 1975; Conner et al. 1982). Since saturated fatty acids, from a qualitative point of view, are unlikely to affect the performance of daphnids, these changes in the fatty acid composition have been neglected.

Sterols are often considered as indispensable structural components of eukaryotic cell membranes. However, there is experimental evidence that tetrahymanol, hopanoids, and other polyterpenoids are functionally equivalent to sterols in stabilizing phospholipid bilayers (Ourisson et al. 1987; Raederstorff and Rohmer 1988). Therefore, one could speculate that also in daphnids, tetrahymanol and related compounds can be incorporated into cell membranes and, to a certain extent, replace sterols as stabilizers of cell membranes. This would explain the observed intermediate food quality of ciliates for *Daphnia* when the ciliates are cultured on prokaryotic food sources (DeBiase et al. 1990; Sanders et al 1996). In a previous study, we provided evidence for the ability of ciliates to biochemically upgrade a sterol-free diet for subsequent use by *Daphnia* (Martin-Creuzburg et al., submitted), most likely by producing tetrahymanol and related compounds that at least partly released *Daphnia* from sterol limitation. Ederington et al. (1995) have already reported the assimilation of tetrahymanol in copepod tissues (mainly in eggs) when ciliates are offered as food, and have suggested that tetrahymanol is functionally equivalent to cholesterol in the crustacean, thereby maintaining minimal egg production. In addition to their role as structural components of cell membranes, sterols also serve as precursors for many bioactive molecules, such as ecdysteroids, which are involved in the process of molting (Goad 1981). Thus, it remains to be tested whether tetrahymanol and related compounds actually improve the performance of crustaceans, possibly by supplementation of these compounds to a sterol-free diet. However, the amounts of tetrahymanol and diplopterol found in the present study were too low to release *Daphnia* from sterol limitation, as indicated by the growth-enhancing effect of sterol supplementation.

Our data clearly show that the food quality of ciliates for *Daphnia* is determined by the sterol availability in the diet of the ciliates. Ciliates prey upon prokaryotic (bacteria, picocyanobacteria) and eukaryotic (picophytoplankton and heterotrophic nanoflagellates) food sources (e.g., Weisse 1993; Jack and Gilbert 1993; Wickham et al. 1993; Šimek et al. 2000; Nakano et al. 2001; Callieri and Stockner 2002). In the former case, ciliates do not have a dietary source of sterols since this lipid class is usually absent in prokaryotes (see the critical review of Volkman 2003). Here we provide evidence that the food quality of ciliates feeding on prokaryotes is comparatively low because of the lack of sterols. However, in the absence of dietary sterols, ciliates produce tetrahymanol and related compounds that might functionally replace sterols as membrane reinforcers in the ciliates and even in *Daphnia*, thereby leading to an upgrading of the prokaryotic food source. Klein Breteler et al. (2004) did not find evidence for trophic upgrading of a sterol-deficient diet by the marine ciliate *Strombidium sulcatum* for copepods, but in this ciliate neither sterols nor tetrahymanol nor other possible sterol surrogates were detected, which might explain its nutritional inadequacy. In contrast, feeding on eukaryotic food sources leads to the accumulation of dietary sterols in ciliates, and thereby to an increase in the food quality of the ciliates, as shown in the present study. Moreover, the predominance of unsuitable sterols is reduced since ciliates, as intermediary grazers, are capable of biochemically upgrading the dietary sterol composition by adding more suitable sterols to the dietary carbon.

Ciliates are abundant protists in freshwater ecosytems, and their importance as a trophic link between picoplanktonic production and higher trophic levels has often been recognized (e.g., Porter et al. 1979; Gifford 1991; Weisse 1993; Callieri and Stockner 2002). However, the data presented here suggest that the carbon transfer efficiency from ciliates to crustacean zooplankton depends on the availability of sterols in the ciliates' food source. This might also explain the observed differences in the food quality of ciliates for metazoan grazers (Sanders and Wickham 1993) and the variability in the proposed significance of ciliates for transferring mass and energy via the microbial loop to higher trophic levels.

Acknowledgments—We thank P. Merkel for excellent technical assistance and M. Rohmer for providing us with standards of tetrahymanol and diplopterol. This work was supported by the German Research Foundation (DFG, EI 179/4-2), and by a post-doc fellowship to A. Bec from the Conseil Régional d'Auvergne, France.

Chapter 7

Concluding remarks and perspectives

In freshwater ecosystems, the zooplankton is often dominated by cladocerans of the genus *Daphnia*; due to their abundance and their high grazing impact on the phytoplankton, they represent a crucial link between primary and secondary production. However, the transfer of mass and energy at the phytoplankton—*Daphnia* interface is frequently constrained by the predominance of phytoplankton unsuitable to support growth and reproduction of the herbivore. In recent years, the biochemical composition of phytoplankton species has been identified as an important parameter that significantly affects zooplankton performance. Thereby, the attention has focused on the role of polyunsaturated fatty acids (PUFAs) as potentially limiting resources. Here, I provide evidence that the dietary sterol content is another important parameter that should be considered when analyzing the carbon transfer efficiency from phytoplankton to higher trophic levels. Cladocerans as well as copepods are incapable of synthesizing sterols de novo and, therefore, must acquire these essential nutrients from their diet. The dietary sterol content depends basically on the phytoplankton species composition, although the sterol content may also vary within a single algal species with growth conditions (Ballantine et al. 1979, Wright et al. 1980). The phytoplankton is generally a diverse group of organisms including prokaryotic and eukaryotic autotrophic species. In contrast to eukaryotes, prokaryotes usually do not contain sterols (Volkman 2003). In aquatic ecosystems, in particular in eutrophic but also in oligo- to mesotrophic lakes, autotrophic prokaryotes are primarily represented by cyanobacteria, which significantly contribute to primary production. In eutrophic lakes, the predominance of cyanobacteria often leads to a decoupling of primary and secondary production, since cyanobacterial carbon is transferred very inefficiently to the herbivorous zooplankton. Consequently, the reduced impact by grazers leads to the accumulation of cyanobacterial biomass, which is in many cases associated with reduced recreational quality of water bodies and hazards to human health and livestock (Carmichael 1994). In a set of laboratory experiments I have shown that the absence of sterols in cyanobacteria is a major constraint that limits *Daphnia* growth (Von Elert et al. 2003). This suggests an important role of sterols in determining the carbon transfer efficiency from cyanobacteria-dominated phytoplankton to higher trophic levels.

In this thesis, I provide further evidence for the significance of dietary sterols for the performance of the keystone herbivore *Daphnia*. As outlined in chapter 2, a shortage in dietary sterols has serious consequences for a variety of life history traits. Somatic and population growth rates, the number of viable offspring, and the probability of survival were significantly reduced with the diminishing availability of sterols. Moreover, an insufficient sterol supply adversely affected the performance of the offspring, which points to strong maternal effects under sterol limitation. Taking all the observed effects into account, this implies a strong impact of dietary sterols on population dynamics of the herbivorous grazer. In times when sterols are in short supply, e.g. during cyanobacterial blooms, daphnids' growth might be limited by the low availability of sterols, which leads to a reduced energy transfer across the plant—animal interface and finally to a decoupling of primary and secondary production.

Crustaceans generally have a simple sterol composition with characteristic high cholesterol levels (Goad 1981). Cholesterol occupies a central position in arthropod physiology; it is an indispensable structural component of cell membranes and serves as precursor for ecdysteroids and other bioactive molecules. Due to the inability to synthesize sterols de novo, arthropods require a dietary source of sterols. Unlike carnivorous arthropods, which are readily supplied with high amounts of cholesterol, herbivorous species cannot rely on a dietary source of cholesterol since this sterol is hardly found above trace concentrations in plant material (Piironen et al. 2000, Moreau et al. 2002). Therefore, herbivorous species have to use the sterols found in their diet to synthesize cholesterol. In most insects and crustaceans examined to date, common phytosterols, such as sitosterol and stigmasterol, are suitable precursors of cholesterol. Data presented in chapter 3 imply that daphnids are also capable of efficiently dealkylating common phytosterols to cholesterol to meet their demands. Supplementation of a sterol-free diet with sitosterol, stigmasterol or ergosterol, a sterol found in most fungi, but also in many green algae (Akihisa et al. 1992, Rzama et al. 1994, Petkov and Kim 1999), significantly enhanced growth and reproduction of *Daphnia galeata*.

Desmosterol and 7-dehydrocholesterol, which both have been identified as intermediates in arthropods sterol metabolism, also improved growth of the herbivore. However, numerous phytosterols are obviously unsuitable to support the performance of arthropods. For instance, sterols unsaturated at position 7 in the sterol nucleus (Δ^7-sterols) were found to be unsuitable to support growth and development of many insects (Behmer and Nes 2003). Klein Breteler et al. (1999) linked unsuccessful ontogenetic development in copepods to the presence of Δ^7-sterols in the the green alga *Dunaliella* sp., which served as a food source. In this thesis, Δ^7-sterols supported growth and reproduction of *Daphnia* to a significantly lower extent than cholesterol and other common phytosterols. A slightly increased cholesterol content of *Daphnia* suggested that lathosterol was incorporated into animal tissues and that the comparatively low growth enhancing effect is caused by an inefficient cholesterol synthesis from lathosterol. It remains to be tested, if such physiological constraints lead to serious ecological consequences. In contrast to cyanobacteria, which generally do not provide any sterols, eukaryotic phytoplankton is characterized by a large set of phytosterols (Nes and McKean 1977, Volkman 2003). Considering the presented data, it appears unlikely that daphnids are equally adapted to all possible sterol profiles that the animals face during the phytoplankton species succession in different freshwater ecosystems. Hence, the predominance of eukaryotic phytoplankton species with an unsuitable sterol pattern might adversely affect the performance of the zooplankton. Further studies have to reveal physiological and ecological consequences of the presence of unsuitable sterols for *Daphnia* growth.

Heterotrophic nanoflagellates (HNFs) and small ciliates are abundant protists with high grazing impact on heterotrophic and autotrophic components of microbial food webs. The phytoplankton of oligo- to mesotrophic lakes is often dominated by autotrophic picoplankton (APP), mostly by unicellular picocyanobacteria (Weisse 1993, Callieri and Stockner 2002). HNFs and small ciliates are considered to be the most important APP grazers, which themselves contribute significantly to zooplankton nutrition, thus transferring mass and energy via the microbial loop to higher trophic levels (Stoecker and McDowell Capuzzo 1990). As outlined before, picocyanobacteria are unsuitable to support the performance of crustaceans, primarily owing to the lack of sterols. However, heterotrophic protists as intermediary grazers might biochemically upgrade the poor food quality of picocyanobacteria for subsequent use by *Daphnia* by the addition of essential lipids to the dietary carbon. In chapter 4, I investigated the capability of the freshwater heterotrophic flagellate

Paraphysomonas sp. to upgrade different APP diets for *Daphnia* by producing sterols and PUFAs. The results showed that the presence of *Paraphysomonas* sp. as an intermediary grazer in a simplified picocyanobacteria→*Daphnia* food chain enhanced *Daphnia* growth and reproduction by the addition of essential lipids. This was the case for both *Synechococcus* strains used in this study, and also for the toxic strain *Microcystis aeruginosa* (PCC7806), which demonstrated that the heterotrophic flagellate is not only capable of trophic upgrading a poor quality food source by producing essential lipids, but also by detoxifying cyanobacterial food. In contrast, *Paraphysomonas* sp. was incapable of further upgrading the nutritional value of the picoeukaryotic green alga *Choricystis minor*, which itself was a fairly good food source for *Daphnia*. However, even though somatic growth on *C. minor* did not differ from growth on *Paraphysomonas* sp. grown on the green alga, the daphnids produced significantly more eggs, when the intermediary grazer *Paraphysomonas* sp. was interposed. The green alga *C. minor* contained comparatively high amounts of sterols, but was deficient in PUFAs that are essential for *Daphnia*. This emphasized the importance of sterols for somatic growth of *Daphnia* and suggested that certain PUFAs are rather needed for reproduction.

In contrast to heterotrophic flagellates, the capability of ciliates to upgrade a poor quality food source for subsequent use by metazoan grazers has received far less attention. In chapter 5, I addressed the question, whether ciliates can serve as a trophic link between cyanobacterial APP- and *Daphnia*-production, thereby upgrading the nutritional value of an APP-food source. Somatic growth of *Daphnia magna* was significantly enhanced when either of the two ciliate species *Colpidium campylum* or *Cyclidium* sp., which themselves fed on either of two *Synechococcus* strains, was offered as food. The results obtained from this simplified experimental food chain (APP→ciliates→*Daphnia*) suggest that predation on ciliates by *Daphnia* gives access to APP-production and provides a linkage of carbon flow to higher trophic levels. The finding that growth of *Daphnia* on *Synechococcus* is primarily constrained by the lack of sterols implies that the observed trophic upgrading of APP food quality by intermediary ciliates is due to the addition of sterols. However, no sterol could be detected in the ciliates which are presumably unable to synthesize sterols de novo (Conner 1968, Harvey and McManus 1991, Harvey et al. 1997, Klein Breteler et al. 2004). Instead, the pentacyclic triterpenoid alcohol tetrahymanol and related compounds were the only polycyclic alcohols which were detected in comparatively high amounts in the ciliates. Sterols are often considered as indispensable structural components of eukaryotic cell membranes. However, experimental evidence exist that tetrahymanol, hopanoids and other polyterpenoids provide functional equivalence to sterols by stabilizing phospholipid bilayers (Ourisson et al. 1987, Raederstorff and Rohmer 1988). Ciliates are the only known eukaryotes capable of synthesizing sterol surrogates such as tetrahymanol and diplopterol. Data presented in chapter 5 suggest that tetrahymanol and related compounds found in the ciliates functionally replaced sterols as membrane reinforcers also in crustacean tissues. Thus, the observed trophic upgrading of picocyanobacterial food quality by intermediary ciliates might be due to the addition of tetrahymanol and hopanoids, which at least partly released *Daphnia* from sterol limitation.

Despite the production of sterol surrogates, ciliates are obviously less nutritious for daphnids than many algae (DeBiase et al. 1990, Wickham et al. 1993, Sanders et al. 1996, Bec et al. 2003). This is corroborated by the data presented in chapter 6, which show that *Colpidium campylum* grown on an undefined bacterial assemblage is a poor quality food source for *Daphnia*. Thereby, only small amounts of tetrahymanol and diplopterol could be detected in the ciliate, which might explain the nutritional inadequacy for *Daphnia*. However, the food quality of *C. campylum* for *Daphnia* was significantly enhanced when the ciliates were previously fed sterol-supplemented albumin beads. Hence, the nutritional value of ciliates for *Daphnia* is determined by the sterol availability in the ciliates diet. Ciliates feeding on a prokaryotic food source cannot rely on a dietary source of sterols and, therefore, are nutritionally inadequate for daphnids, even though the production of tetrahymanol and related compounds might partly compensate for the sterol deficiency. In contrast, ciliates feeding on eukaryotic food sources accumulate dietary sterols, which leads to an increase in ciliates' food quality for *Daphnia*. After the incorporation of dietary sterols, ciliates further metabolize exogenously supplied sterols to various other sterols (Conner 1968, Harvey and McManus 1991, Harvey et al. 1997). In chapter 6, evidence is provided that ciliates as intermediary grazers are capable of biochemically upgrading the dietary sterol composition for daphnids by adding more suitable sterols to the dietary carbon.

Although heterotrophic protists (HNFs, ciliates) as intermediary grazers represent an additional trophic step at which energy is lost by respiration (Pomeroy and Wiebe 1988), our data suggest that HNFs and ciliates provide a crucial link between poor quality primary producers (e.g. picocyanobacteria) and higher trophic levels. The protist cell densities used in our studies to feed *Daphnia* do not necessarily reflect their natural abundance; they were rather designed to investigate the potential of these protists to serve as food for the crustaceans. In the field, HNFs and ciliates are unlikely the sole food source for *Daphnia*, but they might be an important supplement to nutrient-deficient phytoplankton. Further studies have to reveal how ciliates contribute to zooplankton nutrition, in quantity and quality.

The main topic of this thesis was to assess the importance of sterols for the carbon transfer efficiency across the phytoplankton—herbivore interface in aquatic food webs. I provided evidence that the performance of the aquatic keystone genus *Daphnia* depends on the sterol availability in their phytoplanktonic food source. Recently, it has been shown that also copepods, the second important crustacean group, have a dietary need for sterols (Hassett 2004), which emphasizes the general significance of sterols for trophic interactions in aquatic ecosystems. In particular in eutrophic lakes, where cyanobacteria dominate the phytoplankton, the low sterol availability might lead to a decoupling of primary and secondary production with serious consequences for the entire food web. It has to be mentioned that, apart from the decline in sterols, polyunsaturated fatty acids (PUFAs) are also reduced with the accumulation of cyanobacterial biomass. In a correlative field study, Müller-Navarra et al. (2000) found that the low transfer efficiency between primary producers and consumers during cyanobacterial bloom conditions was related to the reduced availability of eicosapentaenoic acid (EPA). However, experimental evidence for the significance of certain PUFAs in determining the food quality of cyanobacteria is controversial (Von Elert and Wolffrom 2001, Ravet et al. 2003). In a previous study we have shown that daphnids are primarily sterol limited when grown on cyanobacteria and that PUFAs become limiting only when the shortage of sterols has been overcome by sterol supplementation (Von Elert et al. 2003). Furthermore, by supplementation of the filamentous cyanobacterium *Anabaena variabilis* with cholesterol, we have shown that the nutritional inadequacy of this frequently bloom-forming cyanobacterium is also caused by a sterol deficiency and not by mechanical interference with the filtering process of the herbivore. Therefore, sterols should be considered as the major food quality constraint when analyzing carbon and matter fluxes across the cyanobacteria—*Daphnia* interface, even though the final proof under field conditions still has to be adduced. Results derived from laboratory experiments are indispensable for determining the requirements of zooplankton species for single biochemical compounds, such as sterols, and provide a first step toward assessing the ecological relevance of these compounds under field conditions.

As outlined before, heterotrophic protists might be particularly important for the energy transfer to higher trophic levels in oligotrophic lakes. In eutrophic lakes, however, protozoans might also contribute significantly to *Daphnia* nutrition or at least serve as an food supplement to nutrient-deficient cyanobacteria, thereby chanelling carbon and nutrients from heterotrophic components of the aquatic food web to the crustacean grazer. These carbon and matter fluxes might be coupled with the addition of essential lipids produced by the protists.

In recent years, considerable progress has been made in defining food quality aspects and their role for zooplankton performance. However, food quality research has focused mainly on phosphorus and polyunsaturated fatty acids as potentially limiting resources. In this thesis I provided data showing that the dietary sterol content and the sterol composition should also be taken into account when analyzing nutritional constraints and their effects on trophic interactions in aquatic food webs.

.

Abstract

The efficiency of carbon transfer from primary producers to higher trophic levels is an important factor that determines the trophic structure of aquatic food webs. A decoupling of primary and secondary production at the phytoplankton—zooplankton interface can be attributed to the nutritional inadequacy of phytoplankton species. Recently, I have shown that the low carbon-transfer efficiency from cyanobacteria to the keystone genus *Daphnia* is caused by the lack of sterols in cyanobacteria. In the present thesis, further evidence is provided for the significance of dietary sterols for the performance of the herbivore *Daphnia*. A shortage in dietary sterols had serious consequences for a variety of life history traits. Somatic and population growth rates, the number of viable offspring, and the probability of survival were significantly reduced with the diminishing availability of sterols. Moreover, an insufficient sterol supply adversely affected the performance of the offspring, which revealed strong maternal effects under sterol limitation. Taking all the observed effects into account, this points to a strong impact of dietary sterols on population dynamics of the herbivorous grazer *Daphnia*.

Cholesterol, which is of major importance in arthropod physiology, is hardly represented in plant material. Instead, eukaryotic phytoplankton contains a great variety of phytosterols that differ from cholesterol in their chemical structure. Therefore, herbivorous grazers have to use the sterols found in their diet to synthesize cholesterol. In this thesis, I showed that dietary sterols differ in their adequacy to support *Daphnia* growth. The results indicated that Δ^5 and $\Delta^{5,7}$ sterols meet the nutritional requirements of the daphnids, while the Δ^7 sterol lathosterol supports growth and reproduction to a significantly lower extent than cholesterol. Dihydrocholesterol (Δ^0) and lanosterol ($\Delta^{8,24}$) did not improve the growth of *Daphnia*, and growth was adversely affected by the Δ^4 sterol allocholesterol. Thus, it is obvious that structural differences of dietary sterols can have pronounced effects on life history traits of *Daphnia*.

In the field, the carbon that is fixed by primary producers, can be modified prior to ingestion by herbivorous crustaceans. I showed here that heterotrophic flagellates and ciliates as intermediary grazers can serve as a trophic link between picocyanobacteria- and *Daphnia*-production, thereby upgrading the nutritional value of a picocyanobacterial food source. For heterotrophic flagellates, this trophic upgrading is primarily due to the addition of sterols to the dietary carbon. In contrast, ciliates presumably lack the ability to synthesize sterols de novo. Instead, they synthesize the pentacyclic triterpenoid alcohol tetrahymanol and related compounds that provide functional equivalents for sterols as stabilizers of cell membranes. Data presented here suggest that these sterol-surrogates replace sterols as membrane reinforcers also in crustacean tissues, leading to the observed trophic upgrading of the sterol-free picocyanobacterial diet by ciliates.

Despite the production of tetrahymanol and related compounds, ciliates were found to be less nutritious for daphnids than many algae. Here, I presented data showing that the ciliates' food quality for *Daphnia* is significantly enhanced when the ciliates were previously fed on sterol-supplemented albumin beads. Hence, the nutritional value of ciliates is determined by the sterol availability in the ciliates' diet. Ciliates feeding on a prokaryotic food source cannot rely on a dietary source of sterols and, therefore, will be nutritionally inadequate for daphnids, even though the production of tetrahymanol and related compounds might partly compensate for the sterol deficiency. In contrast, ciliates feeding on eukaryotic food sources will accumulate dietary sterols, which leads to an increase in ciliates' food quality for *Daphnia*. Beside the incorporation of dietary sterols, ciliates further metabolized exogenously supplied sterols into various other sterols. By supplementation of sterols unsuitable for supporting *Daphnia* growth, I showed that ciliates as intermediary grazers biochemically upgrade unsuitable dietary sterols to sterols appropriate to meet the physiological demands of *Daphnia*.

Zusammenfassung

Die Effizienz des Kohlenstofftransfers von den Primärproduzenten zu höheren trophischen Ebenen ist ein entscheidener Faktor, der die trophische Struktur aquatischer Nahrungsnetze bestimmen kann. Der geringe Nahrungswert einiger Phytoplankter für das herbivore Zooplankton kann zu einer Entkopplung von Primär- und Sekundärproduktion an der Schnittstelle zwischen Phytoplankton und Zooplankton führen. Eigene Experimente haben gezeigt, dass die geringe Transfereffizienz von cyanobakteriellem Kohlenstoff zu Cladoceren der Gattung *Daphnia* auf das Fehlen von Sterolen in Cyanobakterien zurückzuführen ist. Die in der vorliegenden Arbeit vorgestellten Ergebnisse unterstreichen die Bedeutung von Nahrungssterolen für die Fitness von Daphnien. Ein Mangel an Sterolen in der Nahrung hatte ernsthafte Konsequenzen für die verschiedensten ‚life history traits'. Beispielsweise waren das somatische Wachstum, das Populationswachstum, die Anzahl an lebensfähigen Nachkommen und auch die Überlebenswahrscheinlichkeit der Tiere mit abnehmender Sterolverfügbarkeit deutlich reduziert. Ein Sterolmangel im Futter der Eltern hatte außerdem negative Auswirkungen auf die Fitness der Nachkommen, was auf ausgeprägte maternale Effekte bedingt durch die Sterollimitierung schließen lässt. Der Sterolgehalt der Nahrung hat demnach einen großen Einfluss auf die Populationsentwicklung von Daphnien.

Cholesterol, das wichtigste Sterol im tierischen Organismus, ist in pflanzlichem Material kaum vertreten. Eukaryontische Phytoplankter enthalten stattdessen eine Vielzahl an verschiedenen Phytosterolen, die sich in ihrer chemischen Struktur von Cholesterol unterscheiden. Herbivore Zooplankter müssen folglich Nahrungssterole für die Synthese von Cholesterol nutzen. Die Ergebnisse dieser Arbeit zeigen, dass sich Nahrungssterole in unterschiedlichem Maße als Cholesterolvorstufe eignen. Es wurde gezeigt, dass Δ^5- und $\Delta^{5,7}$-Sterole in der Nahrung den Sterolbedarf der Tiere decken, während das Δ^7-Sterol Lathosterol nur eingeschränkt nutzbar ist und zu reduziertem Wachstum der Daphnien führt. Dihydrocholesterol (Δ^0) und Lanosterol ($\Delta^{8,24}$) hatten dagegen keinen Einfluss auf das Wachstum der Daphnien, und das Δ^4-Sterol Allocholesterol wirkte sich negativ auf das Wachstum der Tiere aus. Diese Ergebnisse lassen vermuten, dass strukturelle Unterschiede zwischen den verschiedenen Nahrungssterolen die ‚life history' von Daphnien drastisch beeinflussen können.

Der von den Primärproduzenten fixierte Kohlenstoff kann im Freiland vor der Ingestion durch Zooplankter modifiziert werden. In dieser Arbeit hab ich gezeigt, dass heterotrophe Flagellaten und Ciliaten als zwischengeschaltete trophische Ebene ein Bindeglied zwischen

Picocyanobakterien und Daphnien darstellen, womit eine qualitative Aufwertung des cyanobakteriellen Kohlenstoffs für Daphnien verbunden ist. Im Falle der heterotrophen Flagellaten ist dieses ‚trophic upgrading' in erster Linie auf die Synthese von Sterolen zurückzuführen. Im Gegensatz dazu sind Ciliaten vermutlich nicht in der Lage, Sterole de novo zu synthetisieren. Stattdessen produzieren Ciliaten den pentazyklischen triterpenoiden Alkohol Tetrahymanol und verwandte Substanzen, die funktionell die Rolle von Sterolen als Membrankomponente übernehmen. Die hier vorgestellten Daten deuten darauf hin, dass diese Sterolsurrogate auch in den Membranen von Crustaceen Sterole funktionell ersetzen können, was möglicherweise zu einer Aufwertung der sterolfreien cyanobakteriellen Nahrung durch Ciliaten geführt hat.

Allerdings sind Ciliaten, obwohl sie Tetrahymanol und verwandte Substanzen produzieren, von geringerem Nahrungswert für Daphnien als viele Algen. Die hier präsentierten Daten zeigen, dass die Futterqualität von Ciliaten für Daphnien deutlich zunimmt, wenn die Ciliaten zuvor mit sterol-supplementierten Albuminbeads gefüttert wurden. Folglich ist die Futterqualität von Ciliaten durch die Sterolverfügbarkeit in der Ciliatennahrung bestimmt. Ciliaten, die auf einer prokaryontischen Nahrungsquelle wachsen, finden in ihrer Nahrung keine Sterole vor und sind damit von geringerer Futterqualität für Daphnien, wenn auch die Produktion von Tetrahymanol den Sterolmangel teilweise kompensieren könnte. Ciliaten, die auf einer eukaryontischen Nahrungsquelle wachsen, akkumulieren dagegen die in ihrer Nahrung vorhandenen Sterole in ihren Zellmembranen, wodurch sich ihre Futterqualität für Daphnien erhöht. Neben dem Einbau von Nahrungssterolen in die Zellmembranen können Ciliaten Sterole in ihrem Stoffwechsel verändern. In dieser Arbeit wurde gezeigt, dass Ciliaten dazu in der Lage sind, die Sterolzusammensetzung der Nahrung von Daphnien durch die Bereitstellung von für Daphnien geeignete Sterole biochemisch aufzuwerten.

Literature cited

Ahlgren, G., I. B. Gustafsson, and M. Boberg. 1992. Fatty acid content and chemical composition of freshwater microalgae. J. Phycol. **28**: 37-50.

Ahlgren, G., L. Lundstedt, M. T. Brett, and C. Forsberg. 1990. Lipid composition and food quality of some freshwater phytoplankton for cladoceran zooplankters. J. Plankton Res. **12**: 809–818.

Akihisa, T., T. Hori, H. Suzuki, T. Sakoh, T. Yokota, and T. Tamura. 1992. 24β-methyl-5α-cholest-9-11-en-3β-ol two 24β-alkyl-Δ-5-7-9-11-sterols and other 24β-alkylsterols from *Chlorella vulgaris*. Phytochemistry **31**: 1769-1772.

Arts, M. T. 1998. Lipids in freshwater zooplankton: selected ecological and physiological aspects. In: Arts, M. T., and B. C. Wainman (eds.). Lipids in freshwater ecosystems. p. 71-90.

Arts, M. T., M. S. Evans, and R. D. Robarts. 1992. Seasonal patterns of total and energy reserve lipids of dominant zooplanktonic crustaceans from a hyper-eutrophic lake. Oecologia **90**: 560-571.

Azam, F., T. Fenchel, J. G. Field, J. S. Gray, L. A. Meyer-Reil, and F. Thingstad. 1983. The ecological role of water-column microbes in the sea. Mar. Ecol. Prog. Ser. **10**: 257-263.

Ballantine, J. A., A. Lavis, and R. J. Morris. 1979. Sterols of the phytoplankton—effects of illumination and growth stage. Phytochemistry **18**: 1459-1466.

Beaver, J. R., and T. L. Crisman. 1982. The trophic response of ciliated protozoans in freshwater lakes. Limnol. Oceanogr. **27**: 246-253.

Bec, A., C. Desvilettes, A. Véra, C. Lemarchand, D. Fontvielle, and G. Bourdier. 2003*b*. Nutritional quality of a freshwater heterotrophic flagellate: trophic upgrading of its microalgal diet for daphnia. Aquat. Microb. Ecol. **32**: 203-207.

Bec, A., C. Desvilettes, A. Véra, D. Fontvieille, and G. Bourdier. 2003*a*. Nutritional value of different food sources for the benthic Daphnidae *Simocephalus vetulus*: role of fatty acids. Arch. Hydrobiol. **156**: 145-163.

Becker, C., and M. Boersma. 2003. Resource quality effects on life history of *Daphnia*. Limnol. Oceanogr. **48**: 700-706.

Behmer, S. T., and D. O. Elias. 2000. Sterol metabolic constraints as a factor contributing to the maintenance of diet mixing in grasshoppers (Orthoptera: Acrididae). Physiol. Biochem. Zool. **73**: 219-230.

Behmer, S. T., and R. J. Grebenok. 1998. Impact of dietary sterols on life-history traits of a caterpillar. Physiol. Entomol. **23**: 165-175.

Behmer, S. T., and W. D. Nes .2003. Insect sterol nutrition and physiology: a global overview. Adv. Insect Physiol. **31**: 1-72.

Bratbak, G. 1985. Bacterial biovolume and biomass estimations. Appl. Environ. Microbiol. **49**: 1488-1493.

Brett, M. T. 1993. Resource quality effects on *Daphnia longispina* offspring fitness. J. Plankton. Res. **15**: 403-412.

Brett, M. T., and D. C. Müller-Navarra. 1997. The role of highly unsaturated fatty acids in food web processes. Freshw. Biol. **38**: 483-499.

Brett, M.T., D. C. Müller-Navarra, and S. K. Park. 2000. Empirical analysis of the effect of phosphorus limitation on algal food quality for freshwater zooplankton. Limnol. Oceanogr. **45**: 1564-1575.

Broglio, E., S. H. Jonasdottir, A. Calbet, H. H. Jakobsen, and E. Saiz. 2003. Effect of heterotrophic versus autotrophic food on feeding and reproduction of the calanoid copepod *Acartia tonsa*: relationship with prey fatty acid composition. Aquat. Microb. Ecol. **31**: 267-278.

Callieri, C, and J. G. Stockner. 2002. Freshwater autotrophic picoplankton: a review. J. Limnol. **61**: 1-14.

Carmichael, W. W. 1994. The toxins of cyanobacteria. Sci. Am. **270**: 64-72.

Caron, D. A., and J. C. Goldman. 1990. Protozoan nutrient regeneration. In: Capriulo G. M (ed) Ecology of marine protozoa. New York, Oxford University Press, p 283-306.

Carrick, H. J., G. L. F. Fahnenstiel, E. F. Stoermer, and R. G. Wetzel. 1991. The importance of zooplankton-protozoan trophic couplings in Lake Michigan. Limnol. Oceanogr. **36**: 1335-1345.

Cobelas, M. A., and J. Z. Lechardo. 1988. Lipids in microalgae. A review I. Biochemistry. Grasas y Aceites **40**: 118-145.

Conner, R. L., F. B. Mallory, J. R. Landrey, and C. W. L. Iyengar. 1969. The conversion of cholesterol to $\Delta^{5,7,22}$-cholestatrien-3β-ol by *Tetrahymena pyriformis*. J. Biol. Chem. **244**: 2325-2333.

Conner, R. L., J. R. Landrey, and N. Czarkowski. 1982. The effect of specific sterols on cell size and fatty acid composition of *Tetrahymena pyriformis* W. J. Protozool. **29**: 105-109.

Conner, R. L., J. R. Landrey, C. H. Burns, and F. B. Mallory. 1968. Cholesterol inhibition of pentacyclic triterpenoid biosynthesis in *Tetrahymena pyriformis*. J. Protozool. **15**: 600-605.

Conner, R. L., J. R. Landrey, J. M. Joseph, and W. R. Nes. 1978. The steric requirements for sterol inhibition of tetrahymanol biosynthesis. Lipids **13**: 692-696.

Cook, H. W. 1996. Fatty acid desaturation and chain elongation in eukaryotes, pp. 129-152. In: Vance, D. E., J. E. and Vance (eds) Biochemistry of lipids, lipoproteins and membranes. Elsevier Science, Amsterdam.

Cranwell, P. A., G. H. M. Jaworski, and H. M. Bickley. 1990. Hydrocarbons, sterols, esters and fatty acids in six freshwater chlorophytes. Phytochemistry **29**: 145-151.

De Bernardi, R., and G.Giussani. 1990. Are blue-green algae a suitable food for zooplankton? An overview. Hydrobiologia **200/201**: 29-41.

De Lange, H. J., and M. T. Arts. 1999. Seston composition and the potential for *Daphnia* growth. Aquat. Ecol. **33**: 387-398.

DeBiase, A. E., R. W. Sanders, and K. G. Porter. 1990. Relative nutritional value of ciliate protozoa and algae as food for *Daphnia*. Microb. Ecol. **19**: 199-210.

DeMott, W. R. 1998. Utilization of a cyanobacterium and a phosphorus-deficient green alga as complementary resources by daphnids. Ecology **79**: 2463-2481.

DeMott, W. R., and D. C. Müller-Navarra. 1997. The importance of highly unsaturated fatty acids in zooplankton nutrition: evidence from experiments with *Daphnia*, a cyanobacterium and lipid emulsions. Freshw. Biol. **38**: 649-664.

Desvilettes, C., G. Bourdier, C. Amblard, and B. Barth. 1997. Use of fatty acids for the assessment of zooplankton grazing on bacteria, protozoan and microalgae. Freshw. Biol. **38**: 629-637.

Ducklow H., D. A. Purdie, P. J. Williams, and J. M. Davies. 1986. Bacterioplankton: a sink for carbon in a coastal marine plankton community. Science. **232**: 865-867.

Ederington, M., G. B. McManus, and H. R. Harvey. 1995. Trophic transfer of fatty acids, sterols, and a triterpenoid alcohol between bacteria, a ciliate, and the copepod *Acartia tonsa*. Limnol. Oceanogr. **40**: 860-867.

Elser, J. J., K. Hayakawa, and J. Urabe. 2001. Nutrient limitation reduces food quality for zooplankton: *Daphnia* response to seston phosphorus enrichment. Ecology **82**: 898-903.

Ernst A, G. Sandmann, C. Postius, S. Brass, U. Kenter, and P. Böger. 1991. Cyanobacterial picoplankton from Lake Constance II. Classification of isolates by cell morphology and pigment composition. Bot. Acta **105**: 161-167.

Fahnenstiel, G.L., H. J. Carrick, and R. Iturriaga. 1991. Physiological characteristics and food-web dynamics of *Synechococcus* in Lakes Huron and Michigan. Limnol. Oceanogr. **36**: 219-234.

Ferguson, K. A., F. M. Davis, R. L. Conner, J. R. Landrey, and F. B. Mallory. 1975. Effect of sterol replacement in vivo on the fatty acid composition of *Tetrahymena*. J. Biol. Chem. **250**: 6998-7005.

Fogg, G. E. 1995. Some comments on picoplankton and its importance in the pelagic ecosystem. Aquat. Microb. Ecol. **9**: 33-39.

Gershengorn, M. C., A. R. H. Smith, G. Goulston, L. J. Goad, T. W. Goodwin, and T. H. Haines. 1968. The sterols of *Ochromonas danica* and *Ochromonas malhamensis*. Biochemistry **7**: 1698-1706.

Gifford, D. J. 1991. The protozoan-metazoan trophic link in pelagic ecosystems. J. Protozool. **38**: 81-86.

Gilbert, L. I., R. Rybczynski, and J. T. Warren. 2002. Control and biochemical nature of the ecdysteroidogenic pathway. Annu. Rev. Entomol. **47**: 883-916.

Gladu, P. K., G. W. Patterson, G. H. Wikfors, D. J. Chitwood, and W. R. Lusby. 1990. The occurrence of brassicasterol and epibrassicasterol in the chromophycota. Comp. Biochem. Physiol. B **97**: 491-494.

Gliwicz, Z. M., and C. Guisande. 1992. Family planning in *Daphnia*: resistance to starvation in offspring born to mothers grown at different food levels. Oecologia **91**: 463-467.

Gliwicz, Z. M., and W. Lampert. 1990. Food thresholds in *Daphnia* species in the absence and presence of blue-green filaments. Ecology **71**: 691-702.

Goad, L. J. 1981. Sterol biosynthesis and metabolism in marine invertebrates. Pure Appl. Chem. **51**: 837-852.

Goad, L. J., G. G. Holz Jr., and D. H. Beach. 1983. Identification of (24S)-24-methylcholesta-5,22-dien-3β-ol as the major sterol of a marine cryptophyte and a marine prymnesiophyte. Phytochemistry **22**: 475-476.

Greenberg, A.E., R. R. Trussell, and L. S. Clesceri. 1985. Standard methods for the examination of water and wastewater. American Public Health Association, Washington, DC.

Grieneisen, M. L. 1994. Recent advances in our knowledge of ecdysteroid biosynthesis in insects and crustaceans. Insect Biochem. Molec. Biol. **24**: 115-132.

Gugger, M., C. Lyra, I. Suominen, I. Tsitko, J.-F. Humbert, M. S. Salkinoja-Salonen, and K. Sivonen. 2002. Cellular fatty acids as chemotaxonomic markers of the genera *Anabaena*, *Aphanizomenon*, *Microcystis*, *Nostoc* and *Planktothrix* (cyanobacteria). Int. J. Syt. Evol. Microbiol. **52**: 1007-1015.

Guillard, R. R. 1975. Cultures of phytoplankton for feeding of marine invertebrates. In: Smith, W. L., and M. H. Chanley (eds), Culture of marine invertebrate animals, Plenum Press, New York.

Guisande, C., and Z. M. Gliwicz. 1992. Egg size and clutch size in two *Daphnia* species grown at different food levels. J. Plankton. Res. **14**: 997-1007.

Gulati, R. D., and W. R. DeMott. 1997. The role of food quality for zooplankton: remarks on the state-of-the-art, perspectives and priorities. Freshw. Biol. **38**: 753-768.

Hadas, O., and T. Berman. 1998. Seasonal abundance and vertical distribution of protozoa (flagellates, ciliates) and bacteria in Lake Kinneret, Israel. Aquat. Microb. Ecol. **14**: 161-170.

Hai, T., B. Schneider, J. Schmidt, and G. Adam. 1996. Sterols and triterpenoids from the cyanobacterium *Anabaena hallensis*. Phytochemistry **41**: 1083-1084.

Harrison, K. E. 1990. The role of nutrition in maturation, reproduction and embryonic development of decapod crustaceans: a review. J. Shellfish Res. **9**: 1-28.

Harvey, H. R., and G. B. McManus. 1991. Marine ciliates as a widespread source of tetrahymanol and hopan-3β-ol in sediments. Geochim. Cosmochim. Acta **55**: 3387-3390.

Harvey, H. R., G. Eglinton, S. C. M. O'Hara, and E. D. S. Corner. 1987. Biotransformation and assimilation of dietary lipids by *Calanus* feeding on a dinoflagellate. Geochim. Cosmochim. Acta **51**: 3031-3040.

Harvey, H. R., M. C. Ederington, and G. B. McManus. 1997. Lipid composition of the marine ciliates *Pleuronema* sp. and *Fabrea salina*: shifts in response to change diets. J. Eukaryot. Microbiol. **44**: 189-193.

Hassett, R. P. 2004. Supplementation of a diatom diet with cholesterol can enhance copepod egg-production rates. Limnol. Oceanogr. **49**: 488-494.

Hillebrand, H, C.-D. Dürselen, D. Kirschtel, U. Pollingher, and T. Zohary. 1999. Biovolume calculation for pelagic and benthic microalgae. J. Phycol. **35**: 403-424.

Ikekawa, N. 1985. Structures, biosynthesis and function of sterols in invertebrates, p. 199-230. *In* H. Danielsson and J. Sjovall [eds.], Sterols and bile acids. Elsevier/North Holland Biomedical.

Jack, J. D., and J. J. Gilbert. 1993. Susceptibilities of different-sized ciliates to direct suppression by small and large cladocerans. Freshw. Biol. **29**: 19-29.

Jeong, H. J., K. H. Park, J. S. Kim, H. Kang, C. H. Kim, H. J. Choi, Y. S. Kim, J. Y. Park, and M. G. Park. 2003. Reduction in the toxicity of the dinoflagellate *Gymnodinium catenatum* when fed on by the heterotrophic dinoflagellate *Polykrikos kofoidii*. Aquat. Microb. Ecol. **31**: 307-312.

Jeong, H. J., S. K. Kim, J. S. Kim, S. T. Kim, Y. D. You, and J. Y. Yoon. 2001. Growth and grazing rates of the heterotrophic dinoflagellate *Polykrikos kofoidii* on red-tide and toxic dinoflagellates. J. Eukaryot. Microb. **48**: 298-308.

Jungmann, D. 1992. Toxic compounds isolated from *Microcystis* PCC 7806 that are more active against *Daphnia* than two microcystins. Limnol. Oceanogr. **37**: 1777-1783.

Jürgens, K. 1994. Impact of *Daphnia* on planktonic microbial food webs — a review. Mar. Microb. Food Webs **8**: 295-324.

Jürgens, K., O. Skibbe, and E. Jappesen. 1999. Impact of metazooplankton on the composition and population dynamics of planktonic ciliates in a shallow, hypertrophic lake. Aquat. Microb. Ecol. **17**: 61-75.

Jüttner, F., J. Leonhardt, and S. Möhren. 1983. Environmental factors affecting the formation of mesityloxid, dimethylallylic alcohol and other volatile compounds excreted by *Anabaena cylindrica*. J. Gen. Microbiol. **129**: 407-412.

Kaneshiro, E. S., L. S. Beischel, S. J. Merkel, and D. E. Rhoads. 1979. The fatty acid composition of *Paramecium aurelia* cells and cilia: changes with culture age. J. Protozool. **26**: 147-158.

Klein Breteler, W. C. M., M. Koski, and S. Rampen. 2004. Role of essential lipids in copepod nutrition: no evidence for trophic upgrading of food quality by a marine ciliate. Mar. Ecol. Prog. Ser. **274**: 199-208.

Klein Breteler, W. C. M., N. Schogt, M. Baas, S. Schouten, and G. W. Kraay. 1999. Trophic upgrading of food quality by protozoans enhancing copepod growth: role of essential lipids. Mar. Biol. **135**: 191-198.

Knights, B. A., and C. J. W. Brooks. 1967. Isomers of 24-ethylidenecholesterol: gas chromatographic and mass spectrometric characterization. Phytochemistry **8**: 463-467.

Lachaise, F., G. Carpentier, G. Somme, J. Colardeau, P. and Beydon. 1989. Ecdysteroid synthesis by crab Y-organs. J. Exp. Zool. **252**: 283-292.

Lampert, W. 1977. A field study on the dependence of the fecundity of *Daphnia* sp. on food concentration. Oecologia 36:363–369.

Lampert, W. 1977a. Studies on the carbon balance of *Daphnia pulex* as related to environmental conditions. I. Methodological problems of the use of ^{14}C for the measurement of carbon assimilation. Arch. Hydrobiol. Suppl. **48**: 287-309.

Lampert, W. 1977b. Studies on the carbon balance of *Daphnia pulex* as related to environmental conditions. II. The dependence of carbon assimilation on animal size, temperature, food concentration and diet species. Arch. Hydrobiol. Suppl. **48**: 310-335.

Lampert, W. 1977c. Studies on the carbon balance of *Daphnia pulex* de Geer as related to environmental conditions. IV. Arch. Hydrobiol. Suppl. 48:.361-368.

Lampert, W. 1981. Inhibitory and toxic effects of blue-green algae on *Daphnia*. Rev. Ges. Hydrobiol. **66**: 285-298.

Lampert, W. 1981a. Toxicity of the blue-green *Microcystis aeruginosa*: effective defense mechanism against grazing pressure by *Daphnia*. Verh. Internat. Verein. Limnol. 21: 1436-1440.

Lampert, W. 1987. Feeding and nutrition in *Daphnia*. Mem. Ist. Ital. Idrobiol. **45**: 143-192.

Lampert, W. 1991. The dynamics of *Daphnia* in a shallow lake. Verh. Int. Verein. Limnol. **24**: 795-798.

Lampert, W. 1993. Phenotypic plasticity of the size at first reproduction in *Daphnia*—the importance of maternal size. Ecology **74**: 1455-1466.

Lampert, W., and I. Trubetskova. 1996. Juvenile growth rate as a measure of fitness in *Daphnia*. Funct. Ecol. **10**: 631-635.

Li, W. K. W., D. V. Subba Rao, W. C. Harrison, J. C. Smith, J. J. Cullen, B. Irwin, and T. Platt. 1983. Autotrophic picoplankton in the tropical ocean. Science **219**: 292-295.

Lynch, M. 1989. The life history consequences of resource depression in *Daphnia pulex*. Ecology **70**: 246-256.

Makino, W., J. Urabe, J. J. Elser, and C. Yoshimizu. 2002. Evidence of phosphorus-limited individual and population growth of *Daphnia* in a canadian shield lake. Oïkos **97**: 197-205.

Mallory, F.B., J. T. Gordon, and R. L. Connor. 1963. The isolation of a pentacyclic triterpenoid alcohol from a protozoan. J. Am. Chem. Soc. **85**: 1362-1363.

Marchessault, P., and A. Mazumder. 1997. Grazer and nutrient impacts on epilimnetic ciliate communities. Limnol. Oceanogr. **42**: 893-900.

Martin-Creuzburg, D., A. Bec, and E. von Elert. 2005. Trophic upgrading of picocyanobacterial carbon by ciliates for nutrition of *Daphnia magna*. Aquat. Microb. Ecol. **41**: 271-280.

Martin-Creuzburg, D., A. Wacker, and E. von Elert. 2005. Life history consequences of sterol availability in the aquatic keystone species *Daphnia*. Oecologia **144**: 362-372.

Martin-Creuzburg, D., and E. von Elert. 2004. Impact of 10 dietary sterols on growth and reproduction of *Daphnia galeata*. J. Chem. Ecol. **30**: 483-500.

Menden-Deuer, S., and E. J. Lessard. 2000. Carbon to volume relationships for dinoflagellates, diatoms, and other protist plankton. Limnol. Oceanogr. **45**: 569-579.

Mohr, S., and R. Adrian. 2002. Reproductive success of the rotifer *Brachionus calyciflorus* feeding on ciliates and flagellates of different trophic modes. Freshw. Biol. **47**: 1832-1839.

Monod, J. 1950. La technique de culture continue. Theorie et application. Annales de l'Institut Pasteur **79**: 390-410.

Mulheirn, L. J., D. J. Aberhart, and E. Caspi. 1971. Dehydrogenation of sterols by the protozoan *Tetrahymena pyriformis*. J. Biol. Chem. **246**: 6556-6559.

Müller, H. 1989. The relative importance of different ciliate taxa in the pelagic food web of Lake Constance. Microb. Ecol. **18**: 261-273.

Müller-Navarra, D. C. 1995. Evidence that a highly unsaturated fatty acid limits *Daphnia* growth in nature. Arch. Hydrobiol. **132**: 297-307.

Müller-Navarra, D. C., M. T. Brett, A. M. Liston, and C. R. Goldman. 2000. A highly unsaturated fatty acid predicts carbon transfer between primary producers and consumers. Nature **403**: 74-77.

Nakano, S., P. M. Manage, Y. Nishibe, and Z. Kawabata. 2001. Trophic linkage among heterotrophic nanoflagellates, ciliates and metazoan zooplankton in a hypereutrophic pond. Aquat. Microb. Ecol. **25**: 259-270.

Nes, W. R., and M. L. McKean. 1977. Biochemistry of steroids and other isopentenoids. University Park Press, Baltimore.

Nes, W. R., B. C. Sekula, W. D. Nes, and J. H. Adler. 1978. The functional importance of structural features of ergosterol in yeast. J. Biol. Chem. **253**: 6218-6225.

Nes, W. R., J. M. Joseph, J. R. Landrey, S. Behzadan, and R. L. Conner. 1981. Steric effects at C-20 and C-24 on the metabolism of sterols by *Tetrahymena pyriformis*. J. Lipid Res. **22**: 770-777.

Ourisson, G., M. Rohmer, and K. Poralla. 1987. Prokaryotic hopanoids and other polyterpenoid sterol surrogates. Ann. Rev. Microbiol. **41**: 301-333.

Pace, M. L. 1982. Planktonic ciliates: Their distribution, abundance, and relationship to microbial resources in a monomictic lake. Can. J. Fish. Aquat. Sci. **39**:1106-1116.

Pace, M. L., and E. Funke. 1991. Regulation of planktonic microbial communities by nutrients and herbivores. Ecology **72**: 904-914.

Pace, M. L., and J. D. Orcutt. 1981. The relative importance of protozoans, rotifers, and crustaceans in a planktonic community. Limnol. Oceanogr. **26**: 822-830.

Pace, M. L., K. G. Porter, and Y. S. Feig. 1984. Life history variation within a parthenogenetical population of *Daphnia parvula* (Crustacea: Cladocera). Oecologia **63**: 43-51.

Park, S. K., M. T. Brett, D. C. Müller-Navarra, and C. R. Goldman. 2002. Essential fatty acid content and the phosphorus to carbon ratio in cultured algae as indicators of food quality for *Daphnia*. Freshw. Biol. **47**: 1377–1390.

Pernthaler, J., K. Šimek, B. Sattler, A. Schwarzenbacher, J. Bobkova, and R. Psenner. 1996. Short-term changes of protozoan control on autotrophic picoplankton in an oligo-mesotrophic lake. J. Plankton. Res. **18**: 443-462.

Petkov, G. D., and D. D. Kim. 1999. Sterols of the green alga *Coelastrum*. Algolog. Stud. **130**: 89-92.

Piironen, V., D. Lindsay, T. Miettinen, J. Toivo, A. M. and Lampi. 2000. Plant sterols: biosynthesis, biological function and their importance to human nutrition. J. Sci. Food Agr. **80**: 939-966.

Pomeroy, L. R. 1974. The ocean's food web, a changing paradigm. BioScience. **24**: 499-504.

Pomeroy, L. R., and W. J. Wiebe. 1988. Energetics of microbial food webs. Hydrobiologia **159**: 7-18.

Porter, J. A., K. E. Young, and P. A. Beachy. 1996. Cholesterol modification of hedgehog signaling proteins in animal development. Science **274**: 255-259.

Porter, K. G., and R. McDonough. 1984. The energetic cost of response to blue-green algal filaments by cladocerans. Limnol. Oceanogr. **29**: 365-369.

Porter, K. G., and Y. S. Feig. 1980. The use of DAPI for identifying and counting aquatic microflora. Limnol. Oceanogr. **25**: 943-948.

Porter, K. G., J. D. Orcutt, and J. Gerritsen .1983. Functional response and fitness in a generalist filter feeder, *Daphnia magna* (Cladocera: Crustacea). Ecology **64**: 735-742.

Porter, K. G., M. L. Pace, and J. F. Battey. 1979. Ciliate protozoans as links in freshwater planktonic food chains. Nature **277**: 563-565.

Porter, K.G., E. B. Sherr, B. F. Sherr, M. L. Pace, and R. W. Sanders. 1985. Protozoa in planktonic food webs. J. Protozool. **32**: 409-415.

Prahl, F. G., G. Eglinton, E. D. S. Corner, S. C. M. O'Hara, and T. E. V. Forsberg. 1984. Changes in plant lipids during passage through the gut of *Calanus*. J. Mar. Biol. Ass. U. K. **64**: 317-334.

R 2003, The R Foundation for Statistical Computing, Version 1.8.1.

Raederstorff, D., and M. Rohmer. 1988. Polyterpenoids as cholesterol and tetrahymanol surrogates in the ciliate *Tetrahymena pyriformis*. Biochim. Biophys. Acta **960**: 190-199.

Ravet, J. L., M. T. Brett, and D. C. Müller-Navarra. 2003. A test of the role of polyunsaturated fatty acids in phytoplankton food quality for *Daphnia* using liposome supplementation. Limnol. Oceanogr. **48**: 1938-1947.

Rees, H. H. 1985. Biosynthesis of ecdysone, p. 249-293. *In* G. A. Kerkut and L. I. Gilbert [eds.], Comprehensive insect physiology, biochemistry and pharmacology, Pergamon.

Rezanka, T., O. Vyhnalek, and M. Podojil. 1986. Identification of sterols and alcohols produced by green algae of the genera *Chlorella* and *Scenedesmus* by means of gas chromatography-mass spectrometry. Folia Microbiol. **31**: 44-49.

Rice, W.R. 1989. Analyzing tables of statistical tests. Evolution **43**: 223-225.

Rothhaupt, K. O. 1988. Mechanistic resource competition theory applied to laboratory experiments with zooplankton. Nature **333**: 660-662.

Rudolph, P., E. Spaziani, and W. Wang. 1992. Formation of ecdysteroids by Y-organs of the crab, *Menippe mercenaria*: I. Biosynthesis of 7-dehydrocholesterol in vivo. Gen. Comp. Endocr. **88**: 224-234.

Rudolph, P., and E. Spaziani. 1992. Formation of ecdysteroids by Y-organs of the crab, *Menippe mercenaria*: II. Incorporation of cholesterol into 7-dehydrocholesterol and secretion products in vitro. Gen. Comp. Endocr. **88**: 235-242.

Rzama, A, E. J. Dufourc, and B. Arreguy. 1994. Sterols from green and blue-green algae grown on reused waste water. Phytochemistry **37**: 1625-1628.

Sakwinska, O. 2004. Persistent maternal identity effects on life history traits in *Daphnia*. Oecologia **138**: 379-386.

Sanders, R. W , K. G. Porter, S. J. Bennett, and A. E. DeBiase. 1989. Seasonal patterns of bacterivory by flagellates, ciliates, rotifers, and cladocerans in a freshwater planktonic community. Limnol. Oceanogr. **34**: 673-687.

Sanders, R. W., and S. A. Wickham. 1993. Planktonic protozoa and metazoa: predation, food quality and population control. Mar. Microb. Food Webs **7**: 197-223.

Sanders, R. W., C. E. Williamson, P. L. Stutzman, R. E. Moeller, C. E. Goulden, and R. Aoki-Goldsmith. 1996. Reproductive success of "herbivorous" zooplankton fed algal and nonalgal food resources. Limnol. Oceanogr. **41**: 1295–1305.

Sherr, B. F., and E. B. Sherr. 1984. Role of heterotrophic protozoa in carbon and energy flow in aquatic ecosystems. In: Klug, M. J., and C. A. Redy (eds.). Current perspectives in microbial ecology. Amer. Soc. Microbiol. Washington, p 412-423.

Sherr, B. F., E. B. Sherr, and L. J. Albright. 1987. Bacteria: link or sink? Science **235**: 88-89.

Sherr, E. B., and B. F. Sherr. 1988. Role of microbes in pelagic food webs: a revised concept. Limnol. Oceanogr. **33**: 225-1227.

Sherr, E. B., B. F. Sherr, and G. A. Paffenhöffer. 1986. Phagotrophic protozoa as food for metazoans: a 'missing' trophic link in marine pelagic food webs? Mar. Microb. Food Webs 1: 61-80.

Sherr, E. B., B. F. Sherr, T. Berman, and O. Hadas. 1991. High abundance of picoplankton-ingesting ciliates during late fall in Lake Kinneret 1991. J. Plankton Res. 13: 789-799.

Šimek, K, J. Bobkova, M. Macek, J. Nedoma, and R. Psenner. 1995. Ciliate grazing on picoplankton in a eutrophic reservoir during the summer phytoplankton maximum: a study at the species and community level. Limnol. Oceanogr. 40: 1077-1090.

Šimek, K., K. Jürgens, J. Nedoma, M. Comerma, and J. Armengol. 2000. Ecological role and bacterial grazing of *Halteria* spp.: small freshwater oligotrichs as dominant pelagic ciliate bacterivores. Aquat. Microb. Ecol. 22: 43-56.

Šimek, K., M. Macek, J. Pernthaler, V. Straškrabová, and R. Psenner. 1996. Can freshwater ciliates survive on a diet of picoplankton? J. Plankton Res. 18: 597-613.

Sterner, R. W., and D. O. Hessen. 1994. Algal nutrient limitation and the nutrition of aquatic herbivores. A. Rev. Ecol. Syst. 25: 1-29.

Sterner, R. W., and J. L. Robinson. 1994. Thresholds for growth in *Daphnia magna* with high and low phosphorus diets. Limnol. Oceanogr. 39: 1228-1232.

Sterner, R. W., and K. L Schulz. 1998. Zooplankton nutrition: recent progress and a reality check. Aquat. Ecol. 32: 261-279.

Stich, H-B, and W. Lampert. 1984. Growth and reproduction of migrating and non-migrating *Daphnia* species under simulated food and temperature conditions of diurnal vertical migration. Oecologia 61: 192-196.

Stockner, J. G., and K. S. Shortreed. 1989. Algal picoplancton production and contribution to food webs in oligotrophic British Columbia lakes. Hydrobiologia. 173: 151-166.

Stockner, J. G., and N. J. Antia. 1986. Algal picoplankton from marine and freshwater ecosystems: a multidisciplinary perspective. Can. J. Fish. Aquat. Sci. 43: 2472-2503.

Stoecker, D. K., and J. McDowell Capuzzo. 1990. Predation on Protozoa: its importance to zooplankton. J. Plankton Res. 12: 891-908.

Stoecker, D. K., M. W. Parrow, J. M. Kurkholder, and H. D. Glasgow. 2002. Grazing by microzooplankton on *Pfiesteria piscicida* cultures with different histories of toxicity. Aquat. Microb. Ecol. 28: 79-85.

Sul, D., E. S. Kaneshiro, K. Jayasimhulu, and J. A. Erwin. 2000. Neutral lipids, their fatty acids, and the sterols of the marine ciliated protozoon, *Parauronema acutum*. J. Eukaryot. Microbiol. 47: 373-378.

Sul, D., J. A. Erwin. 1997. The membrane lipids of the marine ciliated protozoan *Parauronema acutum*. Biochim. Biophys. Acta 1345:162-171

Sundbom, M.,and T. Vrede. 1997. Effects of fatty acid and phosphorus content of food on the growth, survival and reproduction of *Daphnia*. Freshw. Biol. 38: 665-674.

Svoboda, J. A. and Thompson, M. J. 1985.. Steroids, pp. 137-175. In: Kerkut, G. A. and Gilbert, L. I. (eds.). Comprehensive insect physiology, biochemistry and pharmacology. Pergamon Press, New York.

Takatsuto, S., N. Kosuga, B. Abe, T. Noguchi, S. Fujioka, and T. Yokota. 1999. Occurrence of potential brassinosteroid precursor steroids in seeds of wheat and foxtail millet. J. Plant Res. 112: 27-33.

Tang, K. W., and M. Taal. 2005. Trophic modification of food quality by heterotrophic protists: species-specific effects on copepod egg production and egg hatching. J. Exp. Mar. Biol. Ecol. **318**: 85-98.

Tang, K. W., H. H. Jakobsen, and A. W. Visser. 2001. *Phaeocystis globosa* (Prymnesiophyceae) and the planktonic food web: feeding, growth, and trophic interactions among grazers. Limnol. Oceanogr. **46**: 1860-1870.

Ten Haven, H. L., M. Rohmer, J. Rullkötter, and J. Bisseret. 1989. Tetrahymanol, the most likely precursor of gammacerane, occurs ubiquitously in marine sediments. Geochim. Cosmochim. Acta **53**: 3073-3079.

Teshima, S. 1971. Bioconversion of β-sitosterol and 24-methylcholesterol to cholesterol in marine crustacea. Comp. Biochem. Phys. B **39**: 815-822.

Teshima, S. and A. Kanazawa, 1971*a*. Sterol compositions of marine crustaceans. Bull. Jap. Soc. Sci. Fish. **37**: 63-67.

Teshima, S. and A. Kanazawa. 1973. Metabolism of desmosterol in the prawn, *Penaeus japonicus*. Mem. Fac. Fish. Kagoshima Univ. **22**: 15-19.

Teshima, S., A. Kanazawa., and H. Sasada. 1983. Nutritional value of dietary cholesterol and other sterols to larval prawn, *Penaeus japonicus* Bate. Aquaculture **31**: 159-167.

Teshima, S., and A. Kanazawa. 1971*b*. Bioconversion of the dietary ergosterol to cholesterol in *Artemia salina*. Comp. Biochem. Phys. B **38**: 603-607.

Tessier, A. J., and N. L. Consolatt. 1989. Variation in offspring size in *Daphnia* and consequences to individual fitness. Oikos **56**: 269-276.

Tezuka, Y. 1974. An experimental study on the food chain among bacteria, *Paramecium* and *Daphnia*. Int. Rev. Ges. Hydrobiol. **59**: 31-37.

Urabe, J., and R. W. Sterner. 2001. Contrasting effects of different types of resource depletion on life history traits in *Daphnia*. Funct. Ecol. **15**: 165-174.

Valcarce, G., J. Florin-Christensen, and C. Nudel. 2000. Isolation of a Δ^7-cholesterol desaturase from *Tetrahymena thermophila*. Appl. Microbiol. Biotechnol. **53**: 591-595.

Venkatesan, M. I. 1989. Tetrahymanol: its widespread occurrence and geochemical significance. Geochim. Cosmochim. Acta **53**: 3095-3101.

Vera, A., C. Desvilettes, A. Bec, and G.Bourdier. 2001. Fatty acid composition of freshwater heterotrophic flagellates: an experimental study. Aquat. Microb. Ecol. **25**: 271-279.

Verity, P. G., C. Y. Roberson, C. R. Tronzo, M. G. Andrews, J. R. Nelson, and M. E. Sieraki. 1992. Relationships between cell volume and the carbon and nitrogen content of marine photosynthetic nanoplankton. Limnol. Oceanogr. **37**: 1434-1446.

Volkman, J. K. 2003. Sterols in microorganisms. Appl. Microbiol. Biotechnol. **60**: 495-506.

Von Elert, E. 2002. Determination of limiting polyunsaturated fatty acids in *Daphnia galeata* using a new method to enrich food algae with single fatty acids. Limnol. Oceanogr. **47**: 1764-1773.

Von Elert, E., and P. Stampfl. 2000. Food quality for *Eudiaptomus gracilis*: the importance of particular highly unsaturated fatty acids. Freshw. Biol. **45**: 189-200.

Von Elert, E., and T. Wolffrom. 2001. Supplementation of cyanobacterial food with polyunsaturated fatty acids does not improve growth of *Daphnia*. Limnol. Oceanogr. **46**: 1552-1558.

Von Elert, E., D. Martin-Creuzburg, and J. R. Le Coz. 2003. Absence of sterols constrains carbon transfer between cyanobacteria and a freshwater herbivore (*Daphnia galeata*). Proc. R. Soc. Lond. B **270**:1209-1214.

Von Liebig, J. 1855. Die Grundsätze der Agrikulturchemie. Vieweg.

Wacker, A., and E. von Elert. 2001. Polyunsaturated fatty acids: evidence for non-substitutable biochemical resources in *Daphnia galeata*. Ecology **82**: 2507-2520.

Weers, P. M. M., and R. D. Gulati. 1997. Effect of the addition of polyunsaturated fatty acids to the diet on growth and fecundity of *Daphnia galeata*. Freshw. Biol. **38**: 721-729.

Weers, P. M. M., and R. D. Gulati. 1997. Growth and reproduction of *Daphnia galeata* in response to changes in fatty acids, phosphorus, and nitrogen in *Chlamydomonas reinhardtii*. Limnol. Oceanogr. **42**: 1584-1589.

Weers, P. M. M., K. Siewertsen, and R. D. Gulati .1997. Is the fatty acid composition of *Daphnia galeata* determined by the fatty acid composition of the ingested diet? Freshw. Biol. **38**: 731-738.

Weisse T, H. Müller, R. M. Pinto-Coelho, A. Schweizer, D. Springmann, and G. Baldringer. 1990. Response of the microbial loop to the phytoplankton spring bloom in a large prealpine lake. Limnol Oceanogr **35**: 781-794.

Weisse, T. 1993. Dynamics of autotrophic picoplankton in marine and freshwater ecosystems. In: Jones J. G. (ed) Advances in Microbial Ecology, Vol. 13, Plenum Press, New York, pp. 327-370.

Wickham, S. A., and J. J. Gilbert. 1991. Relative vulnerabilities of natural rotifer and ciliate communities to cladocerans: laboratory and field experiments. Freshw. Biol. **26**: 77-86.

Wickham, S. A., J. J. Gilbert, and U.-G. Berninger. 1993. Effects of rotifers and ciliates on the growth and survival of *Daphnia*. J. Plankton Res. **15**: 317-334.

Wright, D. C., L. R. Berg, and G. W. Patterson. 1980. Effect of cultural conditions on the sterols and fatty acids of green algae. Phytochemistry **19**: 783-785.

Yasuda, S. 1973. Sterol compositions of crustaceans. I. Marine and fresh-water decapods. Comp. Biochem. Phys. B **44**: 41-46.

Yeagle, P. L., R. B. Martin, A. K. Lala, H.-K. Lin, and K. Bloch. 1977. Differential effects of cholesterol and lanosterol on artificial membranes. PNAS **74**: 4924-4926.

Zhukova, N. V., and V. I. Kharlamenko. 1999. Sources of essential fatty acids in the marine microbial loop. Aquat. Microb. Ecol. **17**: 153-157.

Zöllner E, B. Santer, M. Boersma, H.-G. Hoppe, and K. Jürgens. 2003. Cascading predation effects of *Daphnia* and copepods on microbial food web components. Freshw. Biol. **48**: 2174-2193.

Nomenclature of sterols mentioned in the text. $\Delta^{x,y,z}$ designates the number (e.g., 3) and the position of the double bonds (e.g., C_x, C_y, C_z).

Trivial name	Alternative name	IUPAC name	Formula	
Dihydrocholesterol	Cholestan-3β-ol	5α-Cholestan-3β-ol	$C_{27}H_{48}O$	Δ^0
Allocholesterol	Cholest-4-en-3β-ol	Cholest-4-en-3β-ol	$C_{27}H_{46}O$	Δ^4
Cholesterol	Cholest-5-en-3β-ol	Cholest-5-en-3β-ol	$C_{27}H_{46}O$	Δ^5
Lathosterol	Cholest-7-en-3β-ol	5α-Cholest-7-en-3β-ol	$C_{27}H_{46}O$	Δ^7
Zymostenol	Cholest-8-en-3β-ol	Cholest-8(9)-en-3β-ol	$C_{27}H_{46}O$	Δ^8
7-Dehydrocholesterol	Cholesta-5,7-dien-3β-ol	Cholesta-5,7-dien-3β-ol	$C_{27}H_{44}O$	$\Delta^{5,7}$
22-Dehydrocholesterol	Cholesta-5,22-dien-3β-ol	(22E)-Cholesta-5,22-dien-3β-ol	$C_{27}H_{44}O$	$\Delta^{5,22}$
Desmosterol	Cholesta-5,24-dien-3β-ol	Cholesta-5,24-dien-3β-ol	$C_{27}H_{44}O$	$\Delta^{5,24}$
—	Cholesta-7,22-dien-3β-ol	(22E)-Cholesta-7,22-dien-3β-ol	$C_{27}H_{44}O$	$\Delta^{7,22}$
—	Cholesta-5,7,22-trien-3β-ol	(22E)-Cholesta-5,7,22-dien-3β-ol	$C_{27}H_{42}O$	$\Delta^{5,7,22}$
Campesterol	24α-Methylcholest-5-en-3β-ol	Campest-5-en-3β-ol	$C_{28}H_{48}O$	Δ^5
Fungisterol	24β-Methylcholest-7-en-3β-ol	5α-Ergost-7-en-3β-ol	$C_{28}H_{48}O$	Δ^7
Epibrassicasterol	24α-Methylcholesta-5,22-dien-3β-ol	(22E)-Campesta-5,22-dien-3β-ol	$C_{28}H_{46}O$	$\Delta^{5,22}$
Brassicasterol	24β-Methylcholesta-5,22-dien-3β-ol	(22E)-Ergosta-5,22-dien-3β-ol	$C_{28}H_{46}O$	$\Delta^{5,22}$
Ergosterol	24β-Methylcholesta-5,7,22-trien-3β-ol	(22E)-Ergosta-5,7,22-trien-3β-ol	$C_{28}H_{44}O$	$\Delta^{5,7,22}$
Sitosterol	24α-Ethylcholest-5-en-3β-ol	Stigmast-5-en-3β-ol	$C_{29}H_{50}O$	Δ^5
Stigmasterol	24α-Ethylcholesta-5,22-dien-3β-ol	(24E)-Stigmasta-5,22-dien-3β-ol	$C_{29}H_{48}O$	$\Delta^{5,22}$
Poriferasterol	24β-Ethylcholesta-5,22-dien-3β-ol	(24E)-Poriferasta-5,22-dien-3β-ol	$C_{29}H_{48}O$	$\Delta^{5,22}$
Chondrillasterol	24β-Ethylcholesta-7,22-dien-3β-ol	(24E)-5α-Poriferasta-7,22-dien-3β-ol	$C_{29}H_{48}O$	$\Delta^{7,22}$
22-Dihydrochondrillasterol	24β-Ethylcholesta-7-en-3β-ol	5α-Poriferast-7-en-3β-ol	$C_{29}H_{50}O$	Δ^7
Fucosterol	24E-Ethylidenecholesta-5,24(28)-dien-3β-ol	(24E)-Stigmasta-5,24(28)-dien-3β-ol	$C_{29}H_{48}O$	$\Delta^{5,24}$
Isofucosterol	24Z-Ethylidenecholesta-5,24(28)-dien-3β-ol	(24Z)-Stigmasta-5,24(28)-dien-3β-ol	$C_{29}H_{48}O$	$\Delta^{5,24}$
Lanosterol	4,4,14α-Trimethylcholesta-8,24-dien-3β-ol	5α-Lanosta-8,24-dien-3β-ol	$C_{30}H_{50}O$	$\Delta^{8,24}$

Record of achievement / Abgrenzung der Eigenleistung

Chapter 2, 3, 5 and 6:   Results described in these chapters were exclusively performed by myself or under my direct supervision.

Chapter 4:   I contributed to a high degree to the survey and to the processing of the data.

Danke

Ich möchte all denjenigen ganz herzlich danken, die mich bei dieser Arbeit unterstützt haben. Mein ganz besonderer Dank gilt meinem Betreuer Dr. Eric von Elert, der mir jederzeit mit Rat und Tat zur Seite stand. Danke auch für die unzähligen, oft auch kritischen Diskussionen, die wesentlich zum Gelingen dieser Arbeit beigetragen haben. Eric von Elert hat diese Arbeit, die von der Deutschen Forschungsgemeinschaft gefördert wurde (DFG, EI 179/4-2), überhaupt erst möglich gemacht.

Prof. Dr. Karl-Otto Rothhaupt möchte ich für seine wissenschaftlichen Ratschläge und sein Engagement als Gutachter dieser Arbeit danken.

Mein Dank geht auch an Dr. Alexander Wacker für seine fachlichen Ratschläge, für die vielen Diskussionen und auch für die moralische Unterstützung.

Mein Zimmergenosse Patrick Fink hat ganz erheblich zu einer schönen Zeit im Limnologischen Institut beigetragen. Danke für die unzähligen fachlichen und auch privaten Gespräche bei einer Tasse Kaffee. Astrid Löffler, danke für die vielen Spaziergänge und alles andere. Lars Peters, danke für die Hilfe bei EDV-Problemen.

Dr. Alexandre Bec danke ich für das know-how zur Kultivierung von Protozoen und für die experimentelle Unterstützung.

Danken möchte ich außerdem allen weiteren Menschen am Limnologischen Institut. Ganz besonders danke ich Petra Merkel für die exzellente Unterstützung bei der Fettsäure- und Sterolanalytik und für die unzähligen gemessenen Kohlenstoffproben. Danke auch an Christine Gebauer für die Unterstützung im Laboralltag und für die zahlreichen Phosphatanalysen. Frau Beese danke ich für die Bereitstellung des Daphnienfutters. Martin Wolf und Jürgen Gans-Thomsen, vielen Dank für die technische Unterstützung. Frau Huppertz danke ich für ihre Hilfe bei bürokratischen Angelegenheiten.

Zu guter Letzt möchte ich meinem Vater danken, ohne dessen Unterstützung diese Arbeit nie möglich gewesen wäre. Außerdem: Danke an alle, die mich während dieser Arbeit moralisch unterstützt und motiviert haben.

Vielen Dank!

Curriculum Vitae

Name	Dominik Martin-Creuzburg
Day of birth	Mai 17, 1974
Place of birth	Freiburg i. Br.
Nationality	German

Scientific career

2002–2005 Doctoral thesis entitled 'Sterols in *Daphnia* nutrition: physiological and ecological consequences' at the Limnological Institute, University of Konstanz, supervisor PD Dr. Eric von Elert

2003 and 2004 Research at the Leibniz-Institute of Freshwater Ecology and Inland Fisheries, Department of Limnology of Stratified Lakes, Neuglobsow

2001 Practical course at the Bundesanstalt für Gewässerkunde (BfG), Koblenz

2001 Diploma thesis entitled 'Untersuchungen zur Futterqualität von Cyanobakterien: Der Einfluss von Fettsäuren und Sterolen auf das Wachstum von *Daphnia galeata*' at the Limnological Institute, University of Konstanz, supervision Prof. Dr. Karl-Otto Rothhaupt and PD Dr. Eric von Elert

University education

2001 Diploma-degree at the Institute of Biology I (Zoology), University of Freiburg

1998–2001 Studies of Limnology, Zoology and Microbiology at the Universities of Freiburg and Konstanz

1998 Vordiplom degree in biology, University of Freiburg

1995–1998 Basic studies of biology, University of Freiburg

List of Publications

Von Elert, E., Martin-Creuzburg, D. and Le Coz, J. R. (2003). Absence of sterols constrains carbon transfer between cyanobacteria and a freshwater herbivore (*Daphnia galeata*). Proceedings of the Royal Society of London – Series B: Biological Sciences, 270: 1209-1214.

Martin-Creuzburg, D. and Von Elert, E. (2004). Impact of 10 dietary sterols on growth and reproduction of *Daphnia galeta*. Journal of Chemical Ecology, 30: 483-500.

Martin-Creuzburg, D., Wacker, A. and Von Elert, E. (2005). Life history consequences of sterol availability in the aquatic keystone species *Daphnia*. Oecologia, 144: 362-372.

Martin-Creuzburg, D., Bec, A. and Von Elert, E. (2005). Trophic upgrading of picocyanobacterial carbon by ciliates for nutrition of *Daphnia magna*. Aquatic Microbial Ecology, 41: 271-280.

Martin-Creuzburg, D., Bec, A. and Von Elert, E.. Food quality of ciliates for *Daphnia*: the role of sterols. *Submitted*.

Bec, A., Martin-Creuzburg, D. and Von Elert, E. (2006). Trophic upgrading of autotrophic picoplankton food quality by the freshwater heterotrophic flagellate *Paraphysomonas* sp. Limnology and Oceanography, *in press*.

Martin-Creuzburg, D. und Von Elert, E. (2001). Untersuchungen zur Futterqualität von Cyanobakterien: Der Einfluss von Fettsäuren und Sterolen auf das Wachstum von *Daphnia*. Tagungsbericht der Deutschen Gesellschaft für Limnologie, 632-637.

Martin-Creuzburg, D. und Von Elert, E. (2002). Futterqualitätsunterschiede im Phytoplankton: Die Bedeutung von Sterolen für das Wachstum von Daphnien. Tagungsbericht der Deutschen Gesellschaft für Limnologie, 280-285.

Martin-Creuzburg, D. und Von Elert, E (2003). Einfluss der Futterqualität auf life-history traits von *Daphnia*. Tagungsbericht der Deutschen Gesellschaft für Limnologie, 405-409.

Martin-Creuzburg, D., Bec, A. und Von Elert, E. (2004). Biochemische Aspekte der Futterqualität von Protozoen für *Daphnia*: Trophic upgrading von autotrophem Picoplankton durch Ciliaten? Tagungsbericht der Deutschen Gesellschaft für Limnologie, 431-435.